Gurbachan S. Miglani, PhD

Dictionary of Plant Genetics and Molecular Biology

Pre-publication REVIEWS, COMMENTARIES, EVALUATIONS . . .

"This book is a wonderfully complete, valuable, and timely contribution. The definitions of terms are concise, making the book highly readable for scientists or anyone working or interested in the fields of genetics and molecular biology."

Professor Lianzheng Wang
Chairman, Academic Committee of The Chinese Academy of Agricultural Sciences; President of The China Crop Sciences Society, Beijing, China

"This is a timely publication. Our knowledge of genetics, particularly molecular genetics, is progressing at a phenomenal rate. Numerous terms are being used in publications. This dictionary will help students of plant genetics and molecular biology as well as the general public to grasp the meaning of these terms. I congratulate Dr. Miglani on his careful compilation of this dictionary."

M. S. Swaminathan
Chairman, M. S. Swaminathan Research Foundation, Madras, India

NOTES FOR PROFESSIONAL LIBRARIANS AND LIBRARY USERS

This is an original book title published by The Food Products Press, an imprint of The Haworth Press, Inc. Unless otherwise noted in specific chapters with attribution, materials in this book have not been previously published elsewhere in any format or language.

CONSERVATION AND PRESERVATION NOTES

All books published by The Haworth Press, Inc. and its imprints are printed on certified PH neutral, acid free book grade paper. This paper meets the minimum requirements of American National Standard for Information Sciences–Permanence of Paper for Printed Material, ANSI Z39.48-1984.

Dictionary of Plant Genetics and Molecular Biology

THE FOOD PRODUCTS PRESS

Crop Science

Amarjit S. Basra, PhD

Senior Editor

Dictionary of Plant Genetics and Molecular Biology by Gurbachan S. Miglani

---

Forthcoming and New Books from Haworth Agriculture and Horticulture

Horticulture as Therapy: Principles and Practice by Sharon P. Simson and Martha C. Straus

Crop Sciences: Recent Advances by Amarjit S. Basra

Small Fruits in the Home Garden edited by Robert E. Gough and E. Barclay Poling

A Produce Reference Guide to Fruits and Vegetables from Around the World: Nature's Harvest by Donald D. Heaton

Agriculture Health and Safety: Recent Advances by Kelly J. Donham and Risto Rautainen

The Trans-Oceanic Marketing Channel: A New Tool for Understanding Tropical Africa's Export Agriculture by H. Laurens van der Laan

Effects of Grain Marketing Systems on Grain Production: A Comparative Study of China and India by Zhang-Yue Zhou

Sustainable Forests: Global Challenges and Local Solutions edited by O. Thomas Bouman and David G. Brand

Diagnostic Techniques for Improving Crop Production by Benjamin Wolf

Plant Alkaloids: A Guide to Their Discovery and Distribution by Robert F. Raffauf

Dictionary of Plant Genetics and Molecular Biology

Gurbachan S. Miglani, PhD

The Food Products Press
An Imprint of The Haworth Press, Inc.
New York • London

Published by

The Food Products Press, an imprint of The Haworth Press, Inc., 10 Alice Street, Binghamton, NY 13904-1580

Cover design by Monica L. Seifert.

Library of Congress Cataloging-in-Publication Data

Miglani, Gurbachan S.
Dictionary of plant genetics and molecular biology / Gurbachan S. Miglani.
p. cm.
Includes bibliographical references and index.
ISBN 1-56022-871-7 (alk. paper)
1. Plant genetics–Dictionaries. 2. Plant molecular biology–Dictionaries. I. Title.
QK981.M54 1998
581.3′5′03–dc21

97-20226
CIP

ABOUT THE AUTHOR

Gurbachan S. Miglani, PhD, is Associate Professor of Genetics at the Punjab Agricultural University in Ludhiana, India, with over twenty-five years of experience teaching introductory and advanced genetics in India and the United States, including courses in classical genetics, human genetics, molecular genetics, immunogenetics, biochemical genetics, mutagenesis, and evolutionary genetics. Dr. Miglani is the author of more than eighty research papers and four laboratory manuals. A Fellow of the Indian Society of Genetics and Plant Breeding and a member of the Genetics Society of America, Dr. Miglani received his Master's degree in Genetics from Punjab Agricultural University and his PhD from Howard University in Washington, DC.

Preface

Students, teachers, and research workers sometimes find it very difficult to know the precise meaning of many terms and to accurately interpret many concepts in plant genetics that form the foundation of the subject matter. The glossaries provided at the end of some of the textbooks are highly inadequate. No concerted attempt has been made to provide a comprehensive compilation of plant genetics terms in one book; therefore, there was an urgent need to compile a glossary of the terms used in different specializations of genetics in one book. For this compilation, I have used many textbooks, review papers, original research papers, and dictionaries to achieve this objective. I am indebted to all these authors for their landmark works.

This compilation may be used as a reference book by all who desire to have clarity about terms used in different branches of plant genetics. This dictionary includes definitions and descriptions of more than 3,500 terms and concepts used in various specializations of plant genetics—molecular genetics, cytogenetics, biochemical genetics, population genetics, mutagenesis, evolutionary genetics, and plant biotechnology. The glossary will meet the needs of students, teachers, and research workers from universities and research institutions all over the world who are involved in using principles of genetics in both conventional and unconventional methods for crop improvement. Even high school students and their teachers will find this compilation useful.

I do not claim to have included all possible terms relevant to plant genetics in this compilation—although it was my effort to do so. The readers are requested to inform me about the terms that were not included but, in their opinion, should have found a place in this dictionary; thus they will be incorporated next time. I am indebted to many of my undergraduate and graduate students of the Punjab Agricultural University, Ludhiana who prompted me to write this dictionary. All my colleagues of this department have contributed directly or indirectly to this work. Without the persistent interest, complete involvement, editorial suggestions, technical guidance, and moral support of Dr. Amarjit S. Basra (botany), completion of this work would not have been possible. The sincere efforts and help by Dr. Kuldip Singh (plant breeding) and Dr. Ravi (cytogenetics) in including many modern terms and recent concepts in this book are greatly appreciated. I record my appreciation for Dr. Ashok Kumar Sharma (biochemical genetics) who selflessly stood by me during my library work.

I appreciate the hard and sincere work put in by Mr. Gagandeep Singh of Rax Computer Education, Ludhiana, for meticulously typing the manuscript.

Blessings of my mother, Smt. Apar Kaur, have always boosted my morale. Patience shown by my wife, Harjit, and my children, Jimmy and Simmi, is praiseworthy.

A word of thanks to my publisher, The Haworth Press, Inc., who has produced this work in excellent format and design.

User's Guide

- Different terms included in this book have been arranged alphabetically (A-Z) on a word-by-word basis.
- Terms that are of limited use, very general in nature, or self-explanatory have been avoided.
- Cross-references have been provided wherever necessary for economizing space, demonstrating interrelationships, and organizing the material in the book in a clear manner. Cross-referred terms are indicated in small capitalized letters after the word "*See*"; thus, information is seldom repeated in this book.
- Alternate names of terms, when known, have been given while defining the term; such names are followed by "Also referred to as . . ." or "Also known as . . .".
- Authors who coined, described, or contributed toward further understanding of a term have been listed wherever possible in parentheses following the entry, and respective publications are mentioned in the bibliography. Important references have also been included in the bibliography.
- In a number of cases, a term used may have been defined by different authors or it may have different meanings in different contexts. For such terms various definitions are indicated by (1), (2), (3), etc.
- Genetics is a fast-developing subject. The definitions or descriptions given here may be modified subsequently. The author, therefore, accepts no responsibility for the legal validity, accuracy, adequacy, and interpretation of the terms given in this book. This is a modest effort to serve the scientific community.
- Errors brought to our notice will be rectified whenever possible.

A I, A II: Abbreviations for the anaphase of the first and second meiotic division, respectively.

ABC exonuclease: A DNA repair enzyme that removes DNA adducts by hydrolyzing phosphodiester bonds on both sides of the modified nucleotides. (Seeberg and Steinum, 1982)

aberrant ratio: Any departure from Mendelian expectations in F_2 generations. (Sprague and McKinney, 1966)

aberration: Deviating structure of chromosome or chromatid arising by fragmentation and usually followed by union of broken ends into new combinations of segments. An extreme morphological variant.

abiogenesis: The origin of life on earth from nonliving matter under reducing atmosphere when only methane, water vapors, ammonia, and hydrogen were present in the environment. This view was given by A. I. Oparin, who gave the theory of "primary abiogenesis," which can be summed up as "ambiogenesis at first and biogenesis ever after." This is also referred to as "chemical evolution of life on earth." (Oparin, 1961)

absorbance: For solutions that absorb light, the relative optical density provides information about concentration and the character of the solution. Nucleic acids absorb light at a wavelength maximally at 260 nm, whereas proteins also absorb maximally at 280 nm. Thus, the protein content in an impure solution of DNA can be measured. Since the optical density of a DNA solution increases when the DNA is melted (the hyperchromic effect), changes in optical density can be used to provide detailed information about DNA complexity, especially in the use of reduction of optical density after melting, as is used to construct cot plots. Also known as *optical density*.

accessory DNA: Surplus DNA present in certain cells or cell stages due to, e.g., gene amplification.

acentric fragment: A chromosome that lacks a centromere.

acquired character: Those features that an organism takes on during its lifetime through the effect of the environment on somatic tissue and that are not transferred to the next generation.

acridine dye: A member of a family of chemical compounds that cause frameshift mutations by intercalating between adjacent base pairs of the double helix.

acrocentric chromosome: The position of the centrosome very close to one end of the chromosome (subterminal) such that one chromosome arm is short and the other much longer. (White, 1945)

activating enzyme: An enzyme that plays a part in translating the genetic code. Each activating enzyme attaches itself to a specific amino acid and also to the appropriate transfer RNA. *See* AMINOACYL-tRNA SYNTHETASE.

activation domain: The part of a transcription factor that is believed to recruit RNA polymerase to the 5′ end of the gene.

activator: A regulatory substance that binds to a controlling element and acts in a positive fashion, stimulating transcription of a structural gene or genes.

activator protein: A protein that can stimulate gene expression.

activator RNA: A component of eukaryotic gene regulation that is the product of an integrator gene. It complexes with a specific receptor gene. (Britten and Davidson, 1969)

activator-dissociator (Ac-Ds) system: Originally reported by McClintock in 1956. Responsible for turning the expression of genes on or off. Ds gene located on chromosome 9 causes breakage at the site of its location. It functions in the presence of another gene, Ac, that is located on any other chromosome. Both Ac and Ds elements are transposable and affect the expression of other genes. The Ac-Ds system has been used for gene transfer in eukaryotes. (McClintock, 1978)

active euchromatin: A fraction of euchromatin that is actively engaged in transcription.

adaptability: Capacity of genotype to master macroenvironmental factors so as to manifest a consistent performance over time (year/season) and space (location/latitude). Multienvironmental trials enable one to define the adaptability of genotype(s). Adaptability refers to the adjustment of genotypes to specific environmental conditions so as to reach a certain level of phenotypic performance.

A station trial (single site/year trial) is the test of relative adaptability of genotypes under the trial. In relation to selection, adaptability represents the property of a genotype that permits its survival under selection. Reproductive fitness is thus analogous to adaptability.

adaptation: (1) The process and the fact that individuals or populations of a species are "fitted" to their environment. In part, adaptation is due to phenotypical plasticity representing individual modifications; in part, it is due to genotypic adaptation through natural selection in heterogeneous populations. *See* ECOTYPE. (2) The process, and the result of the process, by which members of a genetic group, either as individuals, as parts of individuals, or as groups are fitted to past or present changes in their environment.

adaptive compensation: A process of natural selection in which a relative lack of fitness in one character may be compensated by special suitability in another.

adaptive enzyme: An enzyme that is produced in response to a specific environment. See INDUCIBLE ENZYME.

adaptive landscape: In the evolutionary theory of Sewall Wright, a representation of the fitness of a population as a topographic map in which contour height (e.g., the fitness value) is a function of allele frequencies at many loci. (Wright, 1932)

adaptive radiation: Evolutionary divergence of the members of a single major stock into a number of distinctive types.

adaptive trials: Multilocational coordinated trials conducted by breeders, agronomists, and plant protectionists across the crop-growing regions in the country to test varietal adaptation over time and space.

adaptive value: The reproductive success of different genotypes of a population under natural conditions in the same environment.

adaptive zone: A set of ecological niches. A population or taxon makes a transition from one zone to another across an adaptive threshold.

adaptor hypothesis: A hypothesis (now experimentally proved) which states that the amino acid sequence of a protein is determined, during the course of genetic translation, by the alignment of aminoacyl-transfer RNAs at corresponding nucleotide codons in messenger RNA. (Crick, 1958; Hoagland, 1959)

addition haploid: An individual having one alien chromosome more than the genetic chromosome number. It is derived from disomic alien addition line.

additive allelic effects: Genetic factors that raise or lower the value of a phenotype on a linear scale of measurement.

additive fitness (h = 0.5): A selection regime in which the relative fitness of the heterozygote is precisely midway between the fitness of the two homozygotes.

additive gene action: The mode of gene action that causes differences in effect or average character expression associated with the two homozygotes for each gene pair involved. The additive effect is also referred to as *breeding value*. This is fixable.

additive genes: Nonallelic genes affecting the same trait such that each adds to the effect of others in the phenotype.

additive model: A mechanism of quantitative inheritance in which alleles at different loci either add a fixed amount to the phenotype or add nothing.

additive portion of genetic variance (V_A): Genetic variance due to the difference between homozygotes (for any locus). Heritability in the narrow sense is V_A divided by the total variance.

adenine: A purine base found in DNA and RNA.

adjacent-1 segregation: A separation of centromeres during meiosis in a translocation heterozygote such that homologous centromeres pass to opposite poles.

adjacent-2 segregation: A separation of centromeres during meiosis in a translocation heterozygote such that homologous centromeres are pulled to the same pole.

adult plant resistance: Resistance that is ineffective in seedlings but effective in adult stage against a pathogen.

adult plant resistance genes: The genes that confer resistance to a host in adult stages against a pathogen.

agar-agar: A gelatinous polysaccharide obtained from the red alga *Gelidium corneum* and from several other red algae. It is a solidifying agent that mixed with nutrient media (0.6 to 1.0 percent) forms a gel for growing tissue-cultured plants and for other purposes. It ranges in quality from relatively inert to very impure (complex, undefined). Its firmness as a gelling agent is affected by medium pH (it is softer when the medium is more acidic) and salt concentration (it is softer when the medium is more dilute). Agar gels melt at about 100°C but solidify at about 44°C. Also known as *agar* or *agarose*.

aggressive hybridization: Participation of a subspecies in hybrid formation only as male, not as female, or vice versa. The consequence is the reduction in numbers of the species acting as female under specific conditions, which may threaten extinction of the species.

albinism: Lack of chlorophyll in green plants. An individual who lacks pigment is known as an *albino*.

aleurone layer: The external layer of cells in the endosperm of cereals. Aleurone grains are located in this single layer of cells.

alien chromosome addition: Incorporation of an additional chromosome from alien species into the normal complement of host species. *See* MONOSOMIC ALIEN ADDITION LINES (MAALS).

alien chromosome substitution: Substitution of homoeologous chromosome from an alien species into the genome of other species.

alien chromosome-segment substitution: To avoid associated ill-effects of whole set of chromosomal genes, substitution of only a desired segment of an alien chromosome.

A-line: The male sterile parent line in a cross to produce hybrid seed. The A-line is the seed-producing line, commonly used with reference to production of hybrid sorghum, hybrid wheat, etc.

alkylating agents: Chemical mutagens that cause mutation by adding alkyl group at various positions in DNA.

allele: One of two or more alternative forms of a gene that can exist at a single gene locus. Distinguished by their differing effects on the phenotype. (Johannsen, 1909)

allele exchange: A method of gene replacement that can be used to introduce mutations into the genome of an organism.

allele frequency: A measure of the commonness of an allele in a population; the proportion of alleles of that gene in the population that are this specific type.

allele shift: An alteration of allele frequency caused by selection, that can lead in extreme cases to complete loss of an allele depending on its initial frequency. Similar effects may be a consequence of genetic drift.

allele substitution: The establishment of a previously rare or absent allele as the predominant allele in a population.

allele-specific probes: Short, labeled oligonucleotide probes used to distinguish single nucleotide differences in polymerase chain reaction (PCR) products through hybridization such that only a completely complementary probe will bind stably to the target sequence.

allelic complementation: The production of wild-type or near-wild-type phenotype in an organism that carries two different mutant alleles in the *trans*-arrangement.

allelozyne: Structurally distinct forms of an enzyme due to multiple allelism at a single locus. *See* ISOZYMES.

allogamy: *See* CROSS-FERTILIZATION.

allogene: Pearson's term for a recessive allele, as opposed to protogene.

alloiobiogenesis: An alternation of a sexual with an asexual form or, cytologically, the alternation of a haploid with diploid stage; alternation of generations.

allometry: Different growth rates in different parts of the same organism.

allopatric hybridization: Hybridization between incompletely differentiated species in a border zone, owing to the premature breakdown of a geographic barrier. *Compare* SYMPATRIC HYBRIDIZATION.

allopatric speciation: A natural process of speciation involving formation of a new species by sister ecotypes (belonging to the same population) geographically isolated over a long period such that they maintain reproductive isolation even if the geographical barriers are broken. However, in the case of sympatric speciation, new species may be formed by splitting of the same population even without geographic isolation. This may occur more frequently in experimental populations rather than in natural populations.

allopolyhaploid: Haploid derived from an allopolyploid.

allopolyploid: A polyploid having whole chromosome sets from different species.

"all-or-none" trait: A trait that is either present or absent in a given individual.

allosteric effect: Reversible interaction of a small molecule, causing a change in the shape of a protein and a consequent alteration of the interaction of that protein with a third molecule.

allosyndesis: In allopolyploids, the pairing between the genomes of two different species; intergenomic pairing. (Ljungdahl, 1924)

allotetraploid: A polyploid formed by doubling of chromosome number in a dihybrid between two organisms with different genomes or by fusion of diploid gametes of such organisms.

allozygote: A diploid individual in which two genes of a particular locus are not identical by descent from a common ancestor. (McKusick and Ruddle, 1977)

allozymes: Forms of an enzyme, controlled by alleles of the same locus, that differ in electrophoretic mobility. *See* ISOZYMES.

alpha DNA: The clustered repeat sequences of DNA located in the centromeric regions of human chromosomes. *See* DNA HELIX A.

α-helix: A helical configuration taken up by some segments of a polypeptide; one of the most common types of protein secondary structure.

alternate dominance: A theory of sex determination that supposed all individuals to be heterozygous for sex but that the male determiners were dominant in male offspring and the female determiners dominant in female offspring.

alternate segregation: A separation of centromeres during meiosis in a reciprocal translocation heterozygote such that balanced gametes are produced.

alternation of generations: The regularly occurring alternation of haploid (gametophyte) and diploid (sporophyte) phases in the life cycle of sexually reproducing plants. (Hofmeister, 1851)

alternative RNA splicing: Functional mRNA molecules are produced from the same primary transcript by differential removal of introns. (Tamkun, Schwarzbauer, and Hynes, 1984)

altitudinal variation: Variation found among individuals of a species for morphological, biochemical, or molecular trait along altitude.

amber codon: The nonsense codon UAG.

amber mutation: Any change in DNA that creates an amber codon at a site previously occupied by a codon representing an amino acid in a protein.

amber suppressors: The mutant genes that code for tRNAs whose anticodons have been altered so that they can respond to UAG codons as well as—or instead of—to their previous codons.

ambiguous codon: The coding for more than one amino acid by a given codon. *See* GENETIC CODE and CODON.

amDNA: Abbreviation for antimessenger DNA.

ameiosis: Occurrence of one division of meiosis in place of two in a normal meiosis, leading to nonreduction of the mother cell.

Ames test: A procedure that is used to test a substance for its likely ability to cause cancer; the test combines the use of animal tissue to generate active metabolites of the substance with a test for mutagenicity in bacteria.

aminoacyl synthetase: Any one of at least twenty different enzymes that catalyze (1) the reaction of a specific amino acid with ATP to form aminoacyl-AMP (activated amino acids) and pyrophosphate and (2) the transfer of the activated amino acid to sRNA forming aminoacyl-tRNA and free AMP.

aminoacyl-site (A-site): The site on the ribosome to which the aminoacyl-tRNA attaches during translation.

aminoacyl-tRNA: A tRNA molecule covalently bound to an amino acid via an acyl bond between the carboxyl group of the amino acid and the 3′-OH of the tRNA.

aminoacyl-tRNA synthetase: Enzymes that attach amino acids to their proper tRNAs.

amitosis: The division of nucleus without the appearance of chromosomes. (Flemming, 1882)

amorph: A mutation that obliterates gene function; a null mutation. (Muller, 1932)

amorphic alleles: The recessive alleles that are not expressed in either homozygotes or heterozygotes compared to recessive alleles that are not expressed in heterozygotes. These are also known as *silent alleles* or *antimorphic alleles*.

amphibivalent: An interchange ring of four chromosomes.

amphidiploid: An organism produced by the hybridization of two species followed by somatic doubling. It is an allotetraploid that appears to be a normal diploid. (Navashin, 1927)

amphihaploid: An allopolyploid containing the haploid set of chromosomes from each of two species. (Zukov, 1941)

amphikaryon: A nucleus containing two haploid sets of chromosomes.

amphimixis: Reproduction by fusion of two gametes in fertilization. Compare APOMIXIS. (Weismann, 1891)

amphipathic: Structures having two surfaces, one hydrophilic and one hydrophobic. Lipids are amphipathic, and some protein regions may form amphipathic helixes, with one charged face and one neutral face.

amphipolyploid: An allopolyploid resulting due to the doubling of chromosome number in a sterile hybrid derived from hybridization between two unrelated species, one or both of which are polyploid.

amplification: Production of additional copies of a chromosomal sequence (gene) found as intrachromosomal or extrachromosomal DNA. Genetic variability is amplified at various levels.

amplification at the gene level: A mutation at one site produces a new allele. Another mutation at a different site produces another new allele. Intragenic recombination produces another new allele that carries both the mutations. Thus, several unisite mutant alleles, as a result of intragenic recombination, can produce two-site and multisite mutant alleles. Multisite mutant alleles have been reported to be present in natural populations. This is thus a mechanism for amplifying genetic variation at the gene level.

amplification at the genomic level: Variation at gene level gets further amplified at genomic level. Independent assortment, crossing-over, and random union of gametes are the mechanisms for amplification of genetic variation at the genomic level.

amplified fragment length polymorphisms (AFLP): Polymorphic fragments are amplified through polymerase chain reaction (PCR).

anagenesis: The evolutionary process whereby one species evolves into another without any splitting of the phylogenetic tree. *See* CLADOGENESIS. (Rensch, 1947)

analogous structures: Structures that have different origins but superficially resemble in structure as a result of adaptation to similar relationships.

analogue: A substance that corresponds to another in many important respects, but is nevertheless quite distinct; derived from the Greek *ana,* according to, and *logos,* ratio, that is of similar proportions.

analogy: Similarity not due to inheritance from a common ancestor but to similar function.

analytical breeding: In a crop such as the potato, the genetic improvement through crosses and selection at the dihaploid level and then chromosome doubling of dihaploids.

anaphase: The stage of nuclear division characterized by movement of chromosomes from spindle equator to spindle poles. It begins with longitudinal division of the centromeres and closes with the end of poleward movement of chromosomes. (Strasburger, 1884)

anaphase I: The stage of meiosis I characterized by separation of chromosomes of the bivalents from spindle equator to spindle pole. Division of the centromeres does not take place. It closes with the end of poleward movement of chromosomes.

anaphase II: The stage of meiosis II characterized by division of the centromeres and separation of sister chromosomes from spindle equator to spindle pole. It closes with the end of poleward movement of sister chromosomes.

ancient DNA: DNA of organisms that lived long ago and some of which have become extinct. Such DNA is a treasure for understanding molecular evolution.

androdioecy: The condition of plant populations or species in which both male (female sterile) and hermaphrodite individuals are found.

androgamy: Development of embryo neither from egg cell nor from synergid or antipodal cells, but from one of the male gametes itself outside or inside the embryo sac.

androgenesis: Development of plants from male gametophytes.

andromonoecy: The condition of an individual plant or species having individuals that bear both male (female sterile) and hermaphrodite flowers.

aneucentric translocation: A translocation involving the centromere so that an acentric chromosome and a dicentric chromosome result.

aneuhaploid: An individual having one chromosome less or more than haploid chromosome number.

aneuploid (heteroploid): The situation that exists when the nucleus of a cell does not contain an exact multiple of the haploid number of chromosomes; one or more chromosome is present in greater or lesser number than the rest. The chromosomes may or may not show rearrangements.

aneuploidy: Variation in chromosomes number by whole chromosomes, but less than an entire set of chromosomes; e.g., $2n + 1$ (trisomy), $2n - 1$ (monosomy).

angstrom (Å): A measurement of length or distance, often used in describing intra- or intermolecular dimensions, equal to 1×10^{-10} meter and 1×10^{-1} nanometer.

anisogamy: *See* HETEROGAMY.

anisogeny: The property of having different inheritance in reciprocal crosses. Now it is known to be due to cytoplasmic inheritance.

anisoploid: Mixture of diploid, triploid, and tetraploid plants obtained with seed harvested from a mixture of diploid and tetraploid plants.

annealing of DNA: The process whereby two initially separate, but complementary, nucleic acid molecules form a double-stranded structure.

anther culture: The culturing of anthers *in vitro* for the purpose of generating haploid plantlets.

anthesis: The process of dehiscence of the anthers; the period of pollen distribution.

anthocyanin: A soluble glucoside pigment producing either reddish or purplish color in flowers and other parts of plants.

antibiosis: Production by host plant of chemical factors (antibiotic toxic compounds) injurious to invading pests.

anticoding strand: The DNA strand that forms the template for both the transcribed mRNA and the coding strand.

anticodon: The triplet of nucleotides in a tRNA molecule that is complementary to, and base pairs with, a codon in an mRNA molecule.

antileader: A nucleotide sequence in DNA upstream of the initiation codon of a gene that is transcribed but not translated. Antileader generally contains Shine-Dalgarno sequence, which helps in binding of mRNA to the ribosome.

antimessenger DNA: A DNA strand copied from messenger RNA by RNA-dependent DNA polymerase by a process called reverse transcription.

antimutagen: An agent that has the ability to suppress mutagenicity of a mutagen.

antimutator gene: A gene that reduces the normal spontaneous mutation rate. (Drake and Allen, 1968)

antimutator mutations: Mutations of DNA polymerase that decrease the overall mutation rate of a cell or of an organism.

antiparallel strands: A term used to refer to the opposite but parallel arrangement of the two sugar-phosphate strands in double-stranded DNA. The 5′-3′ orientation of one such strand is aligned along the 3′-5′ orientation of the other strand.

antisense DNA: A DNA strand composed of nucleotide sequences complementary to mRNA. This strand is used as a template during transcription.

antisense gene: A gene that is used to synthesize antisense RNA. This is done by reversing the orientation of a gene with respect to its promoter.

antisense oligonucleotides: Antisense nucleotide segments that can be used as potential therapeutics for some infectious diseases and cancer. (Cazenave and Hélène, 1991)

antisense RNA: RNA that is complementary to mRNA, the sense RNA. Antisense and sense RNAs are able to hybridize and play a role in the regulation of some genes. Antisense is used *in vitro* as a probe for hybridization and *in vivo* as a method of blocking gene transcription. (Itoh and Tomizawa, 1980)

antisense strand: The template strand on which the mRNA transcript is assembled. This strand does not have the same sense as the mRNA but is complementary to it.

anti σ factor: Protein that prevents recognition of initiation sites by σ factor of RNA polymerase. An antifactor is synthesized during phage T4 infection.

antitermination factor: *See* ANTITERMINATOR PROTEIN.

antiterminator protein: A protein that, when bound at its normal attachment sites, allows RNA polymerase to read through normal terminator sequences.

antitrailer: A nucleotide sequence in DNA downstream of the termination codon of a gene that is transcribed but not translated. An antileader sequence generally contains a signal sequence for attachment of poly(A) tail to RNA polymerase II transcript.

AP DNA: Depurinated DNA.

AP endonucleases: Endonucleases that initiate excision repair at apurinic and apyrimidinic sites on DNA.

AP sites: Apurinic or apyrimidinic sites resulting from the loss of a purine or pyrimidine residue from the DNA.

apogamy: Development of an embryo sac either from an archisporial cell (generative or haploid apospory) or from the nucellus or other cells (somatic or diploid apospory).

apoinducer: A protein that binds to DNA to initiate transcription by RNA polymerase.

apomixis: Reproduction by seeds containing embryos developed directly from sporophyte cells by a vegetative or asexual process, but not by fertilization.

apoptosis: A physiological mechanism responsible for the process of self-destruction of a cell. There is a controlled autodigestion of the cell through activation of endogenous enzymes. Apoptotic genes have been reported in mammals. Apoptosis plays a role in development, homeostasis, and defense.

aporepressor: A protein product of a bacterial gene that upon interacting with an amino acid forms a repressor with increased affinity for an operator. For example, the *trpR* gene of *Escherichia coli* (*E. coli*) that upon interacting with tryptophan forms a repressor with increased affinity for its operators.

apospory: Development of embryo sac either from an archisporial cell (generative or haploid apospory) or from the nucellus or other cells (somatic or diploid apospory).

apostatic selection: Selection of a rare genotype, e.g., a mutant.

applied genecology: Plant breeding defined in terms of genetics and ecology; genetic response of crop varieties and species to habitat and climate, which has a great bearing on adaptive processes.

artificial chromosomes: Chromosomes constituted by ligating origin of replication, autonomous replicating sequences (ARS), telomeric (TEL), and centromeric (CEN) sequences to any group of genes or DNA.

artificial gene: A double-stranded DNA molecule carrying a specific sequence that will code for a given amino acid sequence and that has been produced *in vitro*. Such a gene may be synthesized using an enzyme technique to form a strand of DNA on an mRNA template using reverse transcriptase and DNA polymerase to convert it into double-stranded DNA. Alternatively, an entirely artificial gene may be created, which will code for a novel amino acid sequence by using a template formed from chemically synthesized oligonucleotides followed by enzymic polymerization.

artificial seeds: Somatic embryos encapsulated in synthetic seed coat. Also known as *synthetic seeds* or *synseeds*.

artificial selection: Choosing by humans, as much as possible, the genotypes contributing to the gene pool of succeeding generations.

asexual reproduction (propagation): Any process of reproduction that does not involve fusion of cells. Vegetative reproduction accomplished by portions of the vegetative body is sometimes distinguished as a separate type of asexual reproduction.

assignment test: A test that determines whether a locus is on a specific chromosome by observation of the concordance of the locus and the specific chromosome in hybrid cell lines. (Ruddle et al., 1972)

assortative mating: The mating of individuals with similar phenotypes.

aster: The region that marks the poles of dividing cells and includes rays of microtubules surrounding a clear area within which two centrioles are located. (Fol, 1877)

asynapsis: Failure of pairing during meiosis between homologous chromosomes.

A+T/G+C ratio: An expression of the relative amount of adenine-thymine pairs to guanine-cytosine pairs in a molecule of DNA.

attachment constriction: Centromere; spindle attachment; insertion region; kinetochore; a nonstaining localized region in each chromosome that remains single or appears to be "attached" to the spindle "fiber" sometime after the rest of the chromosome has divided.

attachment efficiency: The percentage of cells plated (seeded, inoculated) that attach to the surface of the culture vessel within a specified period of time. The conditions under which a percentage determination is made should always be stated.

attenuated strains: Strains of pathogenic microorganisms, mainly bacteria and viruses, that have lost their virulence.

attenuation: A mechanism for controlling gene expression in prokaryotes that involves premature termination of transcription. (Kasai, 1974)

attenuator: A stop site present within an operon which causes premature termination of transcription. (Jackson and Yanofsky, 1973)

attenuator region: A sequence present downstream a gene that provides signals for termination of transcription and attachment of poly(A) tail, thereby regulating gene activity.

attenuator stem: *See* TERMINATOR STEM.

AUG: The codon in messenger RNA that specifies initiation of a polypeptide chain or, within a chain, incorporation of a methionine residue.

augmented block design: An experimental design that is used to test a large number of germplasm lines in a limited area. In this design, germplasm lines are unreplicated and checks (previously tested germplasm lines) are replicated.

autocatalysis: The promotion of a chemical reaction by one of the end products of that reaction.

autogamy: Self-fertilization.

autogenous control: The action of a gene product that either inhibits (negative autogenous control) or activates (positive autogenous control) expression of the gene coding for it.

autonomous: A term applied to any biological unit that can function on its own, that is, without the help of another unit. For example, a transposable element that encodes an enzyme for its own transposition is autonomous.

autonomous controlling element: In maize, it is an active transposon with the ability to transpose. *Compare* NONAUTONOMOUS CONTROLLING ELEMENTS.

autonomous replicating sequence (ARS): The sequence that serves as origin of replication and confers the ability to replicate the linked sequences. Requirements for ARS function include an essential core (25 to 65 bp) that typically contains several copies of an 11 A/T-rich consensus sequence. In addition, flanking regions located 100 to 300 bp from the core consensus sequences are important but not essential to ARS function. (Bouton and Smith, 1986)

autopolyhaploid: Haploid derived from an autopolyploid.

autopolyploid: A polyploid whose sets of chromosomes are all of the same species.

autopolyploidy: A polyploid condition in which the additional chromosome set(s) have been derived from within the same group or species. (Kihara and Ono, 1926)

autoradiograph: A record or a photograph prepared by labeling a substance such as DNA with a radioactive material, such as tritiated thymidine, and allowing the image produced by decay radiations to develop on a film over a period of time.

autosyndesis: Intragenomic pairing in allopolyploids, e.g., pairing of the A with A and the B with B genome. (Ljungdahl, 1924)

autotetraploid: A type of autopolyploid having four copies of each chromosome.

autotriploid: An autopolyploid having three copies of each chromosome.

autozygote: A diploid individual in which the two genes of a locus are identical by descent from an ancestral gene. Homozygosity in which the two alleles are identical by descent, i.e., they are copies of an ancestral gene. This condition is referred to as *autozygosity*.

auxins: A plant growth substance (or plant hormone), natural or synthetic, which effects the elongation of shoots and roots when present at low concentration, but which inhibits growth at higher levels.

axenic culture: A culture without foreign or undesired life forms. An axenic culture may include the purposeful cocultivation of different types of cells, tissues, or organisms.

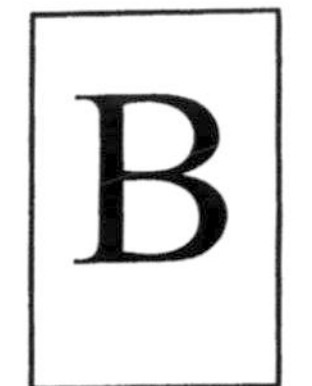

B chromosomes: The chromosomes that are found in addition to the normal chromosomes. Also known as *supernumerary* or *accessory* or *extra chromosomes*. (Randolph, 1928; McClung, 1900)

back cross (BC): The cross of a progeny individual with one of the parents. BC_1, BC_2, etc. are the symbols used to designate the first backcross generation, the second backcross generation, etc.

back mutation: A mutation that reverses the effect of a previous mutation by restoring the original nucleotide sequence.

backcross breeding: The mating of an individual to one of its parents of parental strains.

backcross ratios: The proportion of heterozygotes to bottom recessives expected in backcross in given by calculating 2^n where n is the number of factors involved, e.g., with 2 factors $2^n = 4$ and the expected ratio is 3:1 with 3 factors $2^n = 8$ and expected ratio is 7:1, etc.

bacterial chromosome: A single DNA molecule up to 1 nm long with a molecular weight of over 107 daltons, which may contain around 4,000 to 5,000 kilobases. *See* CHROMOSOME.

bacteriophage: A bacterial virus. Virulent phages induce death in a bacterium. Temperate phages live as prophages in a symbiotic association with bacterium. *See* TEMPERATE PHAGE and PROPHAGE.

balance model of population structure: According to this model, a typical individual is heterozygous at most loci and homozygous at a few loci. This model, which is based on studies with natural populations, is pluralistic in approach and favors existence of a variety of genotypes in the population.

balanced load: That which depresses the overall fitness of a population owing to the segregation of inferior genotypes, the component genes of which are maintained in the population because they add to fitness in different combinations (e.g., as heterozygotes). (Muller, 1950)

balanced polymorphism: A type of genetic polymorphism in which a heterozygote is more adaptive than either homozygote. Natural selection maintains a gene even when it is lethal in homozygous state. In this case, more than one form of a gene is maintained by selection in the same breeding population.

balanced tertiary trisomic: A tertiary trisomic designed in such a way that even if it is heterozygous (Aaa), its trisomic progeny obtained on selfing is genetically similar to the parent. Here the dominant allele is present on the interchanged chromosome linked to the break point.

balancing selection: Heterozygotes when superior to the corresponding homozygotes in gene vigor or some specific component of fitness have a selective advantage. Both alleles will be maintained, and a state of balanced polymorphism will be maintained.

base analog: A chemical compound that is structurally similar to one of the bases in DNA and may act as a mutagen.

base pair substitution: *See* NUCLEOTIDE PAIR SUBSTITUTION.

base pairing rules: The hydrogen-bonded structure formed between two complementary nucleotides (e.g., adenine [A] = thymine [T], and guanine [G] ≡ cytosine [C]). Also known as *competence*. The number of base pairs per turn of the double helix is known as *periodicity*. (Waddington, 1932; Thomas, 1955)

base ratio: The ratio of adenine (A) plus thymine (T) to guanine (G) plus cytosine (C). In DNA the amount of A is equal to the amount of T, and the amount of G is equal to the amount of C, but the amount of A + T does not equal the amount of C + G. The A + T: G + C ratio is constant within a species but varies between species.

base-substitution mutation: A detectable substitution of one or more bases of a nucleotide.

basic chromosome number: The number of chromosomes in ancestral diploid ancestors of polyploids, which is represented by x.

beads-on-a-string: The substructure of chromatin in which individual nucleosomes can be visualized by electron microscopy as beads on a "string" of DNA.

beta particles: Radioactive particles that are generated from radioactive decay of heavier elements such as ^{3}H, ^{32}P, and ^{35}S. These are negatively charged, sparsely ionizing, and more penetrating than nonradioactive particles.

ß-galactosidase: The enzyme that splits lactose into glucose and galactose and encoded by a gene in the *lac* operon.

ß-galactoside acetyl transferase: An enzyme that is involved in lactose metabolism and encoded by a gene in the *lac* operon.

ß-galactoside permease: An enzyme involved in concentrating lactose in the cell and encoded by a gene in the *lac* operon.

ß-sheet: A hydrogen-bonded sheet configuration taken up by some segments of a polypeptide; one of the most common types of protein secondary structure.

Bgl II: A type II restriction from the bacterium *Bacillus flobigii*. The enzyme recognizes the DNA sequence shown below and cuts at the sites indicated by the arrows:

```
     ↓
5′ A G A T C T 3′
3′ T C T A G A 5′
             ↑
```

The sticky ends reproduced by Bgl II are complementary to the ends produced by the enzymes Bam HI, Bcl 1, Xho II, Mbo I, and Sau 3A. Thus, fragments produced by any one of these enzymes will have single-strand extensions that can anneal with the sticky ends on fragments produced by any other of the enzymes above.

bidirectional replication: Type of replication accomplished when two replication forks move away from the same origin in different directions.

bifunctional vector (plasmid): A DNA molecule able to replicate in two different organisms, e.g., in *E. coli* and yeast or *E. coli* and streptomyces. These molecules are thus able to shuttle between the two alternative hosts and are therefore also known as *shuttle vectors*.

bimodal population: A population containing one or more characteristics, the measurements of which show two peaks of relative frequency.

binomial expansion: $(a + b)^n$, where a and b are the probabilities of occurrence and nonoccurrence of an event and must total 1, and n is the number of times the event is tried.

binomial experiment: An experiment with the following properties: (1) experiment repeated under identical conditions (trials); (2) dichotomous outcomes (success and failure); (3) probability [success] = p and is constant from trial to trial; probability [failure] = 1 – p = q; and (4) trials are independent.

bioantimutagens: These are the suppressors of mutagenicity of mutagens, which is accomplished by interfering with the process of mutagenesis. (Kada, Inoue, and Mamiki, 1981)

biochemical genetics: A branch of genetics that deals with the chemical nature of genes and gene action.

biochemical heterosis: A hybrid between two types, each deficient in the agility to synthesize some chemical, grows normally since each type has a normal allele for the deficiency of the other. Consequently, the F_1 hybrid shows hybrid vigor or positive heterosis, in comparison with the parents.

biochemical mutants: Usually bacterial or fungi mutants that cannot grow on minimal media (containing sugar, mineral nutrients, and biotin) unless supplemented by the specific amino acid or vitamin that it is unable to synthesize. They are also called auxotrophs and were first discovered by Beadle in *Neurospora crassa*. Such biochemical mutants have been also reported in some angiosperms, e.g., *Arabidopsis thaliana*.

biochemical mutation: A mutation that alters biochemical function of an individual.

biogenesis: The axiom that life originates only from preexisting life.

biogenetic law: In their development, embryos repeat evolutionary history of their ancestors in an abbreviated form. This is also known as *ontogeny repeats phylogeny*.

biolistic: A method (biological ballistic) of transfecting cells by bombarding them with microprojectiles coated with DNA.

biologic isolation: The isolation of one species from its congeners by reason of interspecfic sterility, incompatibility, preferential mating, or differences in time of sexual activity.

biological containment: One of the precautionary measures taken to prevent replication of recombinant DNA molecules in microorganisms in the natural environment. Biological containment involves the use of vectors and host organisms that have been modified so that they will not survive outside of the laboratory.

biological evolution: Evolution of diverse forms of life from the first simple living cell. It refers to change in diversity and adaptation of populations. Population adapts itself to environment, but the rate of adaptation is slow.

biological property/character/characteristic/trait: Some attributes of individuals within species for which various heritable differences can be defined.

biological race: (1) Such a race may be said to exist where the individuals of a species can be divided into groups usually isolated to some extent by food preferences occurring in the same locality and showing definite differences in biology, but with corresponding structural differences either few, inconstant, or completely absent. (2) In diseases, that the term denotes a strain that, although morphologically indistinguishable, differs from the normal strain in its pathogenicity.

biological species concept: In this species concept, emphasis is placed on "intrinsic" impediments to free exchange of genetic information between populations, i.e., emphasis is on reproductive isolation barriers.

biological yield (Ybiol): Production of the total biomass including economic yield (e.g., grain yield). The larger the Ybiol, the greater the photosynthetic efficiency.

biometrical genetics: A branch of genetics that deals with the inheritance of quantitative traits using statistical concepts and procedures. Also known as *quantitative genetics, statistical genetics,* or *mathematical genetics.*

biometry: The science dealing with the application of statistical methods to biological problems.

biopoiesis: The origin of organisms from replicating molecules. Biopoiesis is a cornerstone of biogenesis. Deoxyribonucleic acid (DNA) is the best example of a self-replicating molecule, and is found in the chromosomes of all higher organisms. In some bacteria, ribonucleic acid (RNA) is self-replicating. Various chemical and physical conditions must be met before either DNA or RNA is able to replicate. *See* ABIOGENESIS.

biotechnology: Commercial applications of useful biological process in industry, medicine, and agriculture.

biotinylated-DNA: A DNA molecule labeled with biotin by incorporation of biotinylated-dUTP into a DNA molecule. It is used as a nonradioactive probe in hybridization experiments such as Southern transfer. The detection of any hybrids uses a complex of streptabidin-biotin-horseradish, which will produce a fluorescent green color when hybrids have formed.

biotype: A population in which all individuals have an identical genotype. (Johannsen, 1903)

biparental mating: Selective or random intermating or segregants in segregating generations (preferably F_2) of a cross. It provides maximum opportunity for recombination leading to rapid release of latent variability (restricted due to linkage disequilibrium) early.

biparental parentage: An individual having two parents, one male and the other female.

biparental zygote: The zygote that contains mitochondrial or chloroplast genes from both the parents. Two alleles for some cytoplasmic genes may thus occur in the zygote. For example, a *Chlamydomonas* zygote that contains cpDNA from both parents; such cells generally are rare.

bipartite: Consisting of two identifiable parts.

bivalent: A pair of synapsed homologous chromosomes. (Haecker, 1892)

bivalent ring: Bivalent with terminal or nearly terminal chiasmata at both ends of the chromosome.

bivalent rod: Bivalent with a terminal or nearly terminal chiasmata formed only at one end.

bivalent, unequal: Bivalent having two partners of unequal size.

blending inheritance: A discredited model of inheritance suggesting that the characteristics of an individual result from the smooth blending of fluid-like influences from its parents.

B-line: The fertile counterpart, or maintainer, of the A-line. The B-line does not have fertility resorter genes and is used as the pollen parent to maintain the A-line. Commonly used with reference to production of hybrid sorghum, hybrid wheat, etc.

block mutation: A term used to denote a change in, or the omission of, a group of adjacent genes.

blocked reading frame: A frame that cannot be translated into protein because it is interrupted by termination codons.

blot: (1) As a verb, this means to transfer DNA, RNA, or protein to an immobilizing matrix such as DBM paper, nitrocellulose, or bio-dyne membranes. (2) As a noun, it usually refers to the autoradio-graph produced during the Southern or Northern blotting procedure.

blunt DNA ends: Ends of a DNA molecule, at which both strands terminate at the same nucleotide position with no single-stranded extension as is done by restriction endonuclease Hae III.

blunt-end ligation: The ligating or attaching of blunt-ended pieces of DNA by T4 DNA ligase. Used in creating hybrid vectors.

bottleneck effect: A brief reduction in size of a population, which usually leads to random genetic drift.

boundary between abiological and biological evolution: Origin of genetic code seems to be a good criterion for distinguishing between living and nonliving organisms.

bouquet stage: Zygotene and pachytene in those organisms in which the chromosomes lie in loops with their ends near one part of the nuclear envelope.

bradytelic evolution: Evolution at a much slower rate than horo-telic evolution. (Simpson, 1944)

branch migration: The process in which a crossover point between two duplexes slides along the duplexes. (Lee, Davis, and Davidson, 1970)

breakage and reunion hypothesis: The general mode by which recombination occurs. DNA duplexes are broken and reunited in a crosswise fashion according to the Holliday model.

"breakage first" hypothesis: The theory that chromosomes can suffer fragmentation after which the fragments may either remain loose, become reattached to restore the original gene arrangement, or unite with other fragments to form new aggregates.

breakage-fusion-bridge cycle: Damage that a dicentric chromosome undergoes through each cell cycle.

breeder seed: Seeds of a released variety multiplied by the breeder to produce foundation seeds.

breeding size: The number of individuals in a population actually involved in the reproduction of a given generation.

breeding system: The organization of mating in a species or smaller group that determines the degree of similarity or difference between gametes that are effective in fertilization.

breeding value: In quantitative genetics, the part of the deviation of an individual phenotype from the population mean that is due to the additive effects of alleles.

bridging cross: A cross made to transfer alleles between two sexually isolated species by first transferring the alleles to an intermediate species that is sexually compatible with both.

broad host range: A term used to describe a plasmid or phage that can replicate in a large nucleic acid.

broad-sense heritability: *See* HERITABILITY IN BROAD SENSE.

buffering genes: A complex of polygenes controlling the expression of a major gene and reducing the variability of its phenotype by toning down its reaction to environmental differences.

bulk breeding: The growing of genetically diverse populations of self-pollinated crops in a bulk plot with or without mass selection, followed by single-plant selection.

bulk population selection: Selection procedure in self-pollinated crops. Segregating populations are propagated as bulks until segregation is virtually ceased, at which time selection is initiated.

buoyant density of DNA: A measure of the density or size of DNA determined by the equilibrium point reached by DNA after density gradient centrifugation.

bypass replication: The replication by eukaryotic DNA polymerases past a DNA lesion, such as UV-induced dimers, with only a brief delay. Bypass replication does not require recombination to circumvent the DNA damage that blocks replication. (Radman et al., 1977)

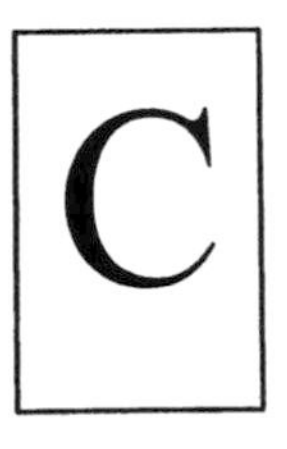

C: Symbol for cytosine in molecular biology and for carbon in chemistry.

C banding: A technique for generating stained regions around centromeres with Giemsa stain using squash technique. The bands constitute heterochromatic regions that are darkly stained. The technique is used extensively in plant cytology.

C value: The amount of DNA found in the haploid genome of a species. (Swift, 1950)

CAAT box: A component of the nucleotide sequence that comprises the eukaryotic promoter. The sequence GGT/CCAATCT that lies between −70 and −80 position is conserved in some promoters (ß-globin gene) but not in others (*TK* gene of herpes simplex virus). (O'Hare, Benoist, and Breathnack, 1981)

callus: An unorganized, proliferative mass of differentiated plant cells; a response to a wound.

callus culture: The cultivation of callus, usually on solidified medium and initiated by inoculation of small explants from established organ cultures or other cultures (inocula). Callus may be maintained indefinitely by regular subdivision of a subculture. It may be used as the basis for organogenetic (shoot, root) cultures, cell cultures, or proliferation of embryoids.

cAMP: *See* CYCLIC AMP.

cap site: A nucleotide sequence upstream of some bacterial genes and operons; the attachment point for the catabolite activator protein. It is the start site of the transcription of a eukaryotic gene.

cap-binding protein: Protein (24 KD) with affinity for cap structure at 5′ end of mRNA that probably assists, together with other initiation factors, in binding the mRNA to the 40S ribosomal subunit. Translation of mRNA *in vitro* is faster if it has a cap-binding protein. (Sonenberg et al., 1978)

capped 5′ ends: The 5′ ends of eukaryotic mRNAs that are modified after transcription to form ends with the general structure 7-methylguanosine-(5′)ppm(5′)2-O-methylated nucleoside.

capping: (1) Concentration of a specific protein at localized region of the surface of a cell. The process is energy-dependent and can be shown, by the use of fluorescent antisera, to commence with generalized binding to all parts of the cell surface, followed by flow of the bound molecules to form a cap or accumulation at one location on the surface. (2) One of the posttranscriptional modifications to much of the hnRNA and mRNA synthesized in eukaryotic cells. Capping consists of the addition of a methylated guanosine residue to the 5′ end of the RNA. *See* TAILING.

cardinal nucleotides: The 3′ nucleotide of an anticodon triplet is considered to be the most important and is called a cardinal nucleotide.

cassette mechanism of mating switch: The mechanism by which homothallic yeast cells alternate mating types. The mechanism involves two silent transposons (cassettes) and a region where these cassettes can be expressed (cassette player).

cassette mutagenesis: A procedure of site-directed mutagenesis that can create specific mutants or families of mutants. Cassette mutagenesis involves removal of a stretch of DNA flanked on either side by a restriction site and insertion of a new cassette of synthetic DNA in its place, which directs various amino acid replacements. (Wells, Vesser, and Powers, 1985)

cat box: A conserved sequence found within the promoter region of the protein-encoding genes of many eukaryotic organisms. It has the canonical sequence GGPyCAATCT and is believed to determine the efficiency of transcription from the promoter.

catabolite activator protein (CAP): A protein that when bound with cyclic AMP can attach to sites on sugar-metabolizing operons to enhance transcription of these operons. (Zubay, Schwartz, and Beckwith, 1970)

catabolite repression: Glucose-mediated reduction in the rates of transcription of operons that specify enzymes involved in catabolic pathways (such as the *lac* operon). (Magasanik, 1961)

catenated DNA: Interlocked circular DNA molecules that are the exclusive products of bacteriophage λ integrative reaction *in vitro* when substrate is a supercoiled molecule containing both attachment sites, *attP* and *attB*. From the Latin *catena*, which means chain.

catenation: The formation of rings or chains by chromosomes at diakinesis.

cauliflower mosaic virus (CaMV): A plant virus, which is a member of the caulimoviruses, that attacks Cruciferae and some Solanaceae. The genome of CaMV consists of a double-stranded, circular DNA molecule containing approximately 8,000 base paris. The circular molecule has several single-stranded discontinuities or gaps occuring at specific locations in relation to the restriction sites, which have been mapped in detail from a number of isolates. The CaMV has been considered as a possible vector for transfer of foreign DNA into plant cells.

caulimoviruses: A group of plant viruses that contain double-stranded DNA. These viruses have a host range restricted to a few closely related plants. They are mainly transmitted by aphids. Infection leads to the formation of refractile round inclusions consisting of many viruses embedded in a protein matrix. The protein is virus-coded and can account for up to 5 percent of the protein-infected cells. The best studied member of this group is the cauliflower mosaic virus (CaMV).

caulogenesis: The process of differentiation of shoots from callus.

cdc genes: Cell division cycle genes, of which several have now been defined, especially in yeasts.

cDNA: *See* COMPLEMENTARY DNA.

cDNA clone: A duplex DNA sequence representing an RNA and carried in a cloning vector.

cDNA cloning: A method of cloning the coding sequence of a gene starting with its mRNA transcript. It is normally used to clone a

DNA copy of a eukaryotic mRNA. The cDNA copy, being a copy of a mature messenger molecule, will not contain any intron sequences and may be readily expressed in any host organism if attached to a suitable promoter sequence within the cloning vector.

cDNA library: A library of clones that have been prepared from mRNA after conversion into double-stranded DNA.

cell: A basic unit of structure and function in all living organisms. (Hooke, 1665)

cell affinity chromatography: A method of obtaining defined, functionally homogenous cell populations from mixed cultures using affinity chromatography. The affinity adsorbents are prepared by immobilizing proteins (antibodies, lectins, etc.) onto macro-beads. The sample is applied to a column packed with this material, impurities are washed out, and the cells are selectively desorbed by washing them in a buffer with excess competitive agent (e.g., free monosaccharides for elution from immobilized lectins).

cell counter: An automated device designed to enumerate the number of cells in a sample. The simpler type of cell counters consist of a mechanism to draw a known volume of a suitably diluted cell culture through a detection device. The detector comprises a plate with a small hole in it, across which an electric potential is maintained. As any cell or particle passes through the hole, a pulse is generated, which is amplified and logged using an electronic counter. More sophisticated cell counters include flow cytometers.

cell culture: Term used to denote the maintenance or cultivation of cells *in vitro* including the cultures of single cells. In cell cultures, the cells are no longer organized into tissues.

cell cycle: The period between one cell division and the next. Divided into M (mitotic), G_1 (gap 1), S (synthetic), and G_2 (gap 2) phases. *See* CELL DIVISION.

cell cycle control: Multiple genetic changes such as loss of fidelity in replication, repair, and genome segregation occur during evolution of normal cells into cancer cells. Fidelity in these processes is normally achieved by the coordinated activity of cyclin-dependent kinase, checkpoint controls, and repair pathways. (Hartwell and Kastan, 1994)

cell disruption: A procedure used to liberate the content of a cell. These may be mechanical and result in cell breakage or depend on cell lysis induced by addition of a solvent that affects the cell membrane, antibiotics, or antimetabolites that disrupt or disorganize cell wall growth.

cell division: The process of reproduction or formation of new cell matter from the preexisting cell in living organisms.

cell generation time: The interval between consecutive divisions of a cell. This interval can best be determined, at present, with the aid of cine photomicrography. This term is not synonymous with *population doubling time.*

cell line: A cell line arises from a primary culture at the time of the first successful subculture. The term "cell line" implies that cultures from it consist of lineages of cells originally present in the primary culture. The terms "finite" or "continuous" are used as prefixes if the status of the culture is known. If not, the term "line" will suffice. The term "continuous line" replaces the term "established line."In any published description of a culture, one must make every attempt to publish the characterization or history of the culture. If such has already been published, a reference to the original publication must be made. In obtaining a culture from another laboratory, the proper designation of the culture, as originally named and described, must be maintained. Any deviation in cultivation from the original must be reported in any publication.

cell plate formation: A process through which a plant cell undergoes cytokinesis.

cell selection: Genetically distinct cells are selected within an individual. (Darlington, 1937a)

cell sorter: A mechanical device designated to separate mixtures of cells. *See* FLOW CYTOMETRY.

cell strain: A cell strain is derived either from a primary culture or a cell line by the selection or cloning of cells having specific properties or markers. In a description of a cell strain, its specific features must be defined. The terms "finite" or "continuous" are to be used as prefixes if the status of the culture is known. If not, the term "line" will suffice.

cell suicide: Death of an active cell in the absence of any internal or external injury. This occurs through a self-destructive, genetically controlled process called "apoptosis." *See* APOPTOSIS.

cell theory: All organisms are wholly constituted of cells. (Schwann, 1838)

cell-free extract: A fluid containing most of the soluble molecules of a cell, made by breaking open cells and getting rid of remaining whole cells.

cell-free system: A mixture of cytoplasmic components from cells, lacking nucleic acids and membranes. Used for *in vitro* protein synthesis and other purposes.

cell-specific gene: *See* LUXURY GENES.

cellular affinity: Tendency of cells to adhere specially to cells of the same type, but not of different types. This property is lost in cancer cells.

cellular oncogenes (C-onc): The cancer genes found in the cells of a host organism.

cellular reprogramming: All phenomena in which the protein pattern of a cell is altered without changes in its DNA.

CEN: A cloned eukaryotic centromere. Each CEN is given a number corresponding to the chromosome from which it was derived. This notation is confined to yeast molecular biology at the moment but, in principle, could be widely applied. CEN has three conserved sequence elements referred to as centromere DNA elements (CDEs). CDEI is a conserved 8 bp palindrome, CDEII is an A/T-rich region of variable block (78 to 86 bp), and CDEIII is a conserved 25 bp sequence with palindromic characteristics.

cenospecies: All the ecospecies that are so related that they are able to exchange genes without loss of fertility or vigor in the offspring.

center of origin: A place where wild-type species are found in greatest genetic variation. (Vavilov, 1922)

centimorgan (cM): The map unit used to describe the distance between two genes on a chromosome. One cM is the distance that corresponds to a 1 percent probability of recombination in a single meiotic event. Its size depends on the genome size of the organism. For example, 1 cM represents 140 kb in *Arabidopsis,* 510 kb in tomatoes, and 1,100 kb in humans.

central body: (1) Centrosome; (2) the structures at the center of the aster during mitosis.

central dogma: The key hypothesis in molecular genetics, proposed by Crick in 1958, which states that DNA makes RNA, which makes protein.

central spindle: A miniature spindle that, where centrosomes and asters are present, may be located between them, and that, on the disappearance of nuclear membrane (at prometaphase), moves into the middle of the nuclear zone. (Hermann, 1891)

centric fragment: A piece of chromosome containing a centromere.

centric fusion: Large arms of two acrocentric chromosomes unite to form a metacentric chromosome. (Robertson, 1963)

centrifugal speciation: A concept of speciation in which isolation is not a factor but is believed to involve a sequential series of mutations that radiate from a central point.

centriole: The central granule in the centrosome. Some experiments suggest the presence of double-stranded DNA in centrioles. (Boveri, 1895)

centrogene: One of the similar and self-propagating units into which the centromere can be broken by the use of X rays or can break itself by misdivision.

centromere: The structural feature of a chromosome, that is, the point at which the pair of chromatids of the metaphase chromosome are held together. (Waldeyer, 1888)

centromere interference: Inhibitory effect exerted by the centromere on crossing-over in the chromosome region immediately adjacent to the centromere.

centromere markers: Loci located near their centromeres.

centromere misdivision: This occurs if a centromere undergoes an anomalous transverse division instead of normal longitudinal division. If a break occurs in the center of centromere, isochromosomes are produced. Breaks in other regions produce telocentric chromosomes.

centromeric DNA: DNA located in the centromeric regions of chromosomes. It consists of highly repeated short sequences of DNA without a gene-coding function, and due to its anomalous structure, it may be recovered as a centromeric satellite fraction on DNA centrifugation. Centromeric DNA is also highly condensed at all times and is therefore referred to as *constitutive heterochromatin*. Also known as *simple sequence DNA*. *See* CENTROMERE.

centromeric fission: Creation of two chromosomes from one by splitting the centromere.

centrosome: A self-propagating cytoplasmic body present in animal cells and some lower plant cells, consisting of a centriole and astral rays at each pole of the spindle during nuclear division.

C-factor: Any gene or heterozygous chromosome structural change that reduces the frequency of crossing-over. (Muller, 1916)

Chambon's rule: The base sequence of an intron begins with GT and ends with AG.

chaperon: A molecular chaperon is a protein that is needed for the assembly or proper folding of some other protein but which is not itself a component of the target complex.

character displacement: Competition exerts a strong selection pressure. In order to reduce competition, natural selection leads to divergence of character. Thus, character displacement takes place through divergent evolution. (Brown and Wilson, 1956)

character divergence: The name given by Darwin to the differences developing in two (or more) related species in their area of sympatry owing to the selective effects of competition. *See* CHARACTER DISPLACEMENT. (Darwin, 1859)

characteristic: An observable or measureable biological attribute of an individual.

Chargaff's rule: In the base composition of DNA, the quantity of adenine equals the quantity of thymine and the quantity of guanine equals the quantity of cytosine (having an equal purine and pyrimidine content).

charged tRNA: A tRNA carrying an amino acid.

charomid: A cosmid vector up to 52 kb in length, bearing 1 to 23 copies of a 2 kb spacer fragment linked in head-to-tail tandem arrays. Like cosmids and lambda phage, charomids can be packaged *in vitro* for efficient introduction into bacteria, and they allow cloning of restriction fragments of any size up to ~45 kb. The number of spacer fragments is variable and determines the lengths of inserts that can be accommodated within the total length of up to 52 kb. (Saito and Stark, 1986)

charon phages: A phage that contains two nonsense mutations which prevent phage from growing in *E. coli* strains that do not carry an appropriate nonsense suppressor mutation. This limits phage growth only to special laboratory bacterial strains and thus acts as a safety feature against the spreading of undesirable strains.

check cross: A cross of an individual of unknown genotype with a phenotypically similar individual of known genotype in order to determine, in F*2,* whether the same gene or allelic series is responsible for the phenotypic appearance of both individuals.

chemical mutagen: One of many chemicals that is capable of inducing mutations.

chemical mutagenesis: The induction of mutations by the use of chemical agents.

chemical sequencing: One of the two main methods currently used in determining primary sequences DNA molecules. Single-stranded DNA is labeled with ^{32}p at one end using polynucleotide kinase and is then cleaved by four separate types of chemical treatments in reactions run in parallel. The chemistry is designed to cleave that part of the DNA strand which is preferentially adjacent to one of the four types of nucleotides and gives rise to populations of fragments ending at all the possible places and all radiolabeled at one end. These fragments are then separated by electrophoresis and autoradiography, and the sequence is deduced from the patterns of bands observed. (Maxam and Gilbert, 1977)

chemiluminescent labeling: A technique used to label DNA probes. Two different probes are constructed that are complementary to adjacent segments of a gene. Each probe carries part of a chemiluminescent system. When brought together, the two probes hybridize in the same region and emit light, which can be detected using a photomultiplier.

chi form: An intermediate structure in recombination between DNA molecules. (Potter and Dressler, 1976)

chi sequence: An octamer that provides a hotspot for RecA-mediated genetic recombination in *E. coli*.

chi (X) site: Sequence of DNA at which the RecBCD protein cleaves one of the strands during recombination.

chiasma (p. chiasmata): A cross-shaped or X-shaped configuration of the chromosomes in a bivalent in prophase of meiosis I, usually the visible result of prior cytological crossing-over. (Janssens, 1909)

chiasma frequency: The average number of chiasmata formed per bivalent, or in any particular bivalent, in a given organism, under given circumstances.

chiasma interference: The more (negative) or less (positive) frequent occurrence of a second chiasma in the vicinity of the first chiasma than expected by chance. (Mather, 1933)

chiasma terminilization: The shifting of a chiasma from its original site toward the end of the chromosome from earliest diplotene to the onset of anaphase I. (Darlington, 1929a)

chiasma theory of pairing: The hypothesis that whenever two chromosomes which have been paired at pachytene remain associated until metaphase, they do so by virtue of the formation of a chiasma or visible exchange of partners amongst their chromatids.

chimera: An association of tissues of different genetic origin and constitution in the same part of an organism. (Winkler, 1907)

chimeric gene: A recombinant gene having regulatory sequence from one gene and coding sequence from another gene. *See* HYBRID GENE.

chimeric plasmid: A bacteriophage or plasmid cloning vector carrying a novel DNA sequence integrated into its DNA.

chimeric vector: Bacteriophage or plasmid cloning vector carrying a novel DNA sequence integrated into its DNA. *See* CHIMERIC PLASMID.

chi-square (χ^2) distribution: The sampling distribution of the chi-square statistic. A family of curves whose shapes depend on degrees of freedom.

chi-square (χ^2) test: A statistical test for testing goodness of fit between expected (on the basis of a gene hypothesis) and observed numbers.

chloroplast aminoacyl-tRNA synthetase: An aminoacyl-tRNA synthetase isolated from chloroplasts that is specific for a chloroplast tRNA.

chloroplast (cp) DNA: DNA presents in chloroplasts. It is circular double-stranded DNA (as in prokaryotes) and it replicates independently of the nuclear DNA by the rolling circle method. Molecular weight equals 55×10^6 to 97×10^6; length equals 419 μm; size equals 120 to 180 kb units. (Chun, Vaughan, and Rech, 1963)

chloroplast genes: Genetic information carried on the DNA contained within the chloroplast. Since the chloroplast are cytoplasmic and during fertilization cytoplasm comes only from the ovum, characteristics coded for in the chloroplasts do not show Mendelian inheritance; characteristics are only passed on through the female line.

chloroplast genome: The genetic information of the chloroplast, which occurs as multiple copies of a circular, double-stranded DNA molecule. Chloroplast genome codes for four rRNAs (23S, 16S, 5S, 4.5S), several ribosomal proteins, tRNAs (>30), RNA polymerase, larger submit of rubisco, some units of proteins of photosystems I and II, ATPase, and NADH dehydrogenase. The most striking feature is the presence of a divided gene of ribosomal protein of smaller subunits (rps 12). The mechanism of replication of chloroplast DNA is semiconservation and involves rolling circle intermediates. Size of the genome ranges 85 to 2,000 kb in green algae and 120 to 160 kb in land plants.

chloroplast mRNA: An mRNA that is only translated on 70S ribosomes.

chloroplast proteins: The soluble proteins and the insoluble lamellar proteins. These are encoded by nuclear genes and by chloroplast genes. The enzyme rubisco is the best studied of the stroma proteins.

chloroplast ribosome: A 70S ribosome found in eukaryotic chloroplasts. Protein synthesis on 70S ribosomes is inhibited by chloramphenicol but not by cycloheximide; the reverse is true for 80S cytoplasmic ribosomes. Chloroplast ribosome dissociate into 50S and 30S subunits.

chloroplast rRNA: The rRNA that consists of various forms of 23S, 16S, 5S, and 4.5S with a guanine/cytosine content of 55.8 percent.

chloroplast tRNA: A tRNA found in the chloroplast. These tRNAs differ in structure from the cytoplasmic tRNA and are coded for by the chloroplast DNA. The structure of these molecules is closer to that of prokaryotic tRNA than eukaryotic cytoplasmic tRNA. The chloroplast tRNA are not aminoacylated by cytoplasmid aminoacyl-tRNA synthetases.

chromatid: One of the two identical longitudinal halves of a chromosome, which shares a common centromere with a sister chromatid; this results from the replication of chromosomes during interphase. (McClung, 1900)

chromatid bridge: A dicentric chromatid with centromeres passing to opposite poles at anaphase. (Smith, 1935)

chromatid conversion: A type of gene conversion that is inferred from the existence of identical sister-spore pairs in a fungal octad that shows a nonMendelian allele ratio.

chromatid interchanges: Interchanges involving breaks in two or more chromatids or in different loci of the same chromatid and reunion in new ways.

chromatid interference: *See* INTERFERENCE, CHROMATID. Mather (1933).

chromatid nondisjunction: The passing of homologous parts of chromatids to the same pole following crossing-over between homologous differential segments in a multiple interchange hybrid.

chromatid tetrad: The four chromatids (two per chromosome) of a bivalent.

chromatin: Originally, the deeply staining material present in the nuclei of cells and corresponding to the chromosomes. Now used more specifically to refer to the structural association between DNA and protein in chromosomes. (Flemming, 1882)

chromatography: A means of separating closely related compounds by allowing a solution or mixture of them to seep through a differentially absorbent material.

chromomere: Small, stainable thickenings arranged linearly along a pachytene or polytene chromosome. (Fol, 1877; Wilson, 1896)

chromonema: Helically coiled, thread-like strands that make up a chromosome.

chromosomal aberration: All types of changes in chromosome structure such as deletions, duplications, inversions, and translocations.

chromosomal chimera: A chimera in which the components do not have the same chromosome number.

chromosomal cycle: All the changes that take place in the chromosomes of an organism from fertilization to gametogenesis.

chromosomal fibers: Fiber bundles connecting the tiny granules with the poles. These chromosomal fibers have by many authors been regarded as intimately involved in chromosomal movement as traction fibers for the chromosomes.

chromosomal fission: When a chromosome splits into two, thereby increasing the number of chromosomes in the karyotype. Fissions are also called *dissociations.*

chromosomal fusion: When two nonhomologous chromosomes fuse into one, thereby reducing the number of chromosomes in the karyotype.

chromosomal indigestion: Incapability of cells to tolerate genomic multiplicity (polyploidy). Excess chromosomes are eliminated from the nucleus.

chromosomal interference: The increase (negative interference) or decrease (positive interference) in likelihood of a second crossover closely adjacent to another. In most organisms, interference increases with decreased distance between crossover. *See* COINCIDENCE. (Muller, 1916)

chromosomal mutation: Segments of chromosomes, whole chromosomes, or even entire sets of chromosomes may be involved in genetic change. Compare GENE MUTATION.

chromosomal polymorphism: A condition where in the same population various chromosomal rearrangements exist simultaneously.

chromosome: A linkage structure consisting of a specific linear arrangement of genes, with or without spacer sequences between them. Chromosomes of eukaryotes and prokaryotes are involved in two main functions: (a) transmission of genetic information from cell to cell and generation to generation; and (b) ordered expression of this information to control cellular functions and development. Chromosomes are autoduplicating structures of differential complexity in eukaryotes and prokaryotes whose number, morphology, and organization are organism-specific characteristics. Chromosomes of eukaryotes are multimolecular, complex structures (consisting of DNA, RNA, proteins, etc.), show a color reaction with basic dyes, and undergo regular changes in the degree of compaction during mitosis and miosis. These are differentiated into chromomeres, heterochromatin, euchromatin, centromeres, and telomeres. Eukaryotic chromosomes are multireplicon structures. *Bacterial chromosomes* exist in the form of a single, circular, double-stranded DNA, which is folded and condensed with the help of specific RNA- and DNA-binding proteins. It is a single replicon. *Viral chromosomes* occur either in the form of DNA or RNA, which may be single- or double-stranded and linear or circular DNA molecular in different types of viruses. Each chromosome is a single replicon. (Waldeyer, 1888)

chromosome arm: One of the two main segments of the chromosome separated by the centromere. (Navashin, 1912)

chromosome banding: The differentially stained sections of chromosomes visible under light or florescence microscope as a result of treatment with various dyes.

chromosome break: Break that scissions the chromosome, generally induced by radiations.

chromosome bridge: Chromosome region between the two centromeres of a dicentric, which at anaphase is being pulled toward both poles of a spindle.

chromosome complement: Sum total of all the chromosomes in an individual. (Darlington, 1932a)

chromosome complex: A set of chromosomes having a given segmental arrangement.

chromosome congression: The movement of chromosomes onto the metaphase plate, especially at the first meiotic division. (Darlington, 1937b)

chromosome conjugation: Side-by-side association of chromosomes in the early prophase of meiosis.

chromosome constriction: An unspiralized segment of a fixed position found in the metaphase chromosome. (Agar, 1911)

chromosome diminution: The loss or expulsion of a part of a chromosome complement at meiosis or mitosis so that a daughter nucleus is formed without this part.

chromosome elimination: Diminution that occurs at meiosis only. (Seiler and Haniel, 1921)

chromosome engineering: The incorporation of alien chromosomes, or parts of alien chromosomes, into a given chromosome set. (Sears, 1972)

chromosome fragmentation: The breakage of chromosomes when acted upon by radiations, for example.

chromosome fusion: Permanent linear attachment of two chromosomes.

chromosome imprinting: The process by which one of the two genetically homologous chromosomes become altered at predetermined sites to function deferentially from its homolog. (Crouse, 1960)

chromosome jumping: A procedure used for the isolation and cloning of an ordered array of widely separated regions of chromosomal DNA. It constitutes a jumping library in which the clones are separated by a jump length ranging from 50 to 150 kb.

chromosome loss: Failure of a chromosome to become incorporated into a daughter nucleus at cell division.

chromosome map: *See* LINKAGE MAP.

chromosome matrix: A chromophobic membrane enclosing the chromosome proper.

chromosome model: The models that explain the organization of chromatin fibers in a chromosome.

chromosome mutation: *See* CHROMOSOMAL ABERRATION.

chromosome painting: Essentially, fluorescent *in situ* hybridization (FISH), with the only difference being that the probe, instead of being a short DNA fragment as in FISH, represents the complete DNA content of a chromosome. Thus, a chromosome paint represents an aggregation of probes from a chromosome that hybridizes to the entire chromosome of origin. (Pinkel et al., 1988)

chromosome rearrangement: A chromosome aberration involving new juxtapositions of chromosome parts.

chromosome scaffold: Basic architectural form after histones have been removed from chromosomes. In this form, revealed by electron microscopy of spread films of such chromatin, long loops of DNA appear to be subtended by a persisting framework of nonhistone proteins. These are loosely referred to as the scaffold proteins. (Adolph et al., 1977)

chromosome segregation: Separation of the members of a pair of homologs in a manner that only one member is present in any postmeiotic nucleus.

chromosome theory of inheritance: The theory, stated in its most convincing form, that genes reside on chromosomes. (Sutton, 1903)

chromosome thread: The protein framework of the chromosome.

chromosome walking: The analysis of a stretch of DNA by the production from it of overlapping restriction fragments, then using each one as a probe for the next. (Bender, Speirer, and Hogness, 1983)

chromosome-mediated gene transfer (CMGT): The use of an isolated metaphase chromosome as a vehicle for the transfer of genes between cultured cells. CMGT provides a method for fine structural mapping of chromosomes. (McBridge and Ozer, 1963)

circular restriction map: A pictorial representation of the position of the sites recognized by various specific restriction enzymes on circular genomes from bacteriophages, viruses, chloroplasts, mitochondria, bacteria, or cyanobacteria.

***cis*-acting locus:** The sequence that affects the activity only of DNA sequences on its own molecule of DNA; this property usually implies that the locus does not code for protein.

***cis*-acting protein:** A protein having the exceptional property of acting only on the molecule of DNA from which it was expressed.

***cis*-configuration:** Placement of at least two dominant pairs of alleles on one chromosome and the recessives on the homologous chromosome. (Haldane, 1942)

***cis*-dominant mutation:** A mutation in the operator that affects regulation of genes of an operator only in *cis*-arrangement. (Epstein and Beckwith, 1968)

***cis*-position pseudoallele:** The location of two closely linked mutant alleles in one chromosome with the two normal alleles in the homologous chromosome.

***cis*-trans complementation test:** A genetic analysis that tests whether two mutations lie in the same or different genes, and that can also provide information on dominance and recessiveness. The test involves introducing the two mutated genes into a single cell, for example, by introducing an F′ plasmid carrying one mutated gene into a recipient bacterium with a chromosomal copy of the second gene. (Lewis, 1951; Benzer, 1957)

cistron: A segment of DNA specifying one polypeptide chain in protein synthesis. Under the concept of a triplet code, one cistron must contain three times as many nucleotide pairs as amino acids in the chain it specifies. (Benzer, 1957)

cladogenesis: The evolutionary process whereby one species splits into two or more species. This is also known as *true speciation. See* ANAGENESIS. (Rensch, 1947)

cladogenetic asymmetry: The "nowhere to go but up" process, whereby a group (clade) originates at a character state (e.g., size) that is physically restricted to expansion mainly in one direction (e.g., a mammal can only get so small before encountering metabolic problems); the descendant radiation is therefore asymmetric in that character state (e.g., larger size). The resulting trend is an increase in variance.

cladogram: Evolutionary lineages represented through a tree-like diagram.

classical genetics: A study of genetics on an organismal rather than cellular level. Also known as *Mendelian genetics*.

classical model of population structure: This model assumes that a typical diploid individual is homozygous at most and heterozygous at a few loci. This model is based on laboratory mutants whose fitness is much lower than that of normal alleles and considers a genotype homozygous for normal alleles at all the loci as the best.

cleavage division: One of the early divisions of the zygote that leads to the formation of an embryo in animals.

cleavage map: *See* RESTRICTION MAP.

cleistogamous: A plant producing inconspicuous flowers that never open and are thus self-pollinated.

clinal selection: Selection that changes gradually along a geographic gradient.

cline: A gradient of variation in measurement of a characteristic of a population. (Huxley, 1939)

clonal propagation: Asexual reproduction of plants that are considered to be genetically uniform and originated from a single individual or explant.

clone: (1) a population of cells or organisms derived from a single cell or common ancestor by mitosis. A clone may not necessarily comprise of homogenous population. (2) A population of recombinant DNA molecules all carrying the same inserted sequence. (Webber, 1903)

clone library: A collection of clones that contain a number of different genes. *See* GENOMIC LIBRARY.

cloned gene: A gene that has been inserted into a cloning vector.

cloning: *See* MOLECULAR CLONING.

cloning efficiency: The percentage of cells plated (seeded, inoculated) that form a clone. One must be certain that the colonies formed arose from single cells in order to properly use this term. *See* COLONY FORMING EFFICIENCY.

cloning vector (vehicle): A DNA molecule, capable of replication in a host organism, into which a gene is inserted to construct a recombinant DNA molecule.

closed continuous culture: A continuous culture in which flow of fresh medium is balanced by outflow of corresponding volume of spent medium. Cells are separated mechanically from outflowing medium and added back to the culture.

closed mating system: A system in which no outside individuals are permitted to introduce gametes into the population.

closed population: A population with no genetic input other than by mutation.

closed promoter complex: The initial complex formed between RNA polymerase and a promoter, in which the double helix is still completely base paired.

closed reading frame: The reading frame contains termination codons that prevent its translation into protein.

cloverleaf tRNA: A two-dimensional representation of the structure of a tRNA molecule.

cluster clone theory: A prokaryote having a single duplicated strand of DNA has several genes. The DNA strand breaks up into two parts containing different genes. Some of these genes are retained in the nucleus, and other DNA segments are converted into mitochondria through the acquisition of separate membranes. Neither the method nor the source of these membranes is explained. (Bogorad, 1975)

coacervate: An aggregate of varying degrees of complexity, formed by the interaction of two or more colloidal suspensions. Also called a *microsphere*. (Oparin, 1961; Fox and Dose, 1972)

coadaptation: A mechanism (balanced combination of genes) that maintains the genetically balanced (allelic) structure of a gene pool such that change in the frequency of one allele does affect other loci, leading to repatternization of the whole gene pool. (Dobzhansky, 1950)

coated vesicle: A vesicle whose membrane has on its surface a layer of protein coating. (Roth and Porter, 1964)

coconversion: The simultaneous correction of two sites during gene conversion.

code: The genetic code.

code dictionary: A listing of the 64 possible codons and their translational meanings is given in Table 1.

code letter: Nucleotides, e.g., A, U, G, C (in RNA) or A, T, G, C (in DNA).

coding capacity: The amount of protein that can be specified by a given DNA or RNA sequence. Estimates of coding capacity usually require assumptions about the absence of introns or of overlapping genes; as a rough guide, 1 megadalton of double-stranded DNA can encode 60 to 70 thousand daltons of protein.

coding ratio: The number of bases in nucleic acids divided by the number of amino acids whose sequences they determine in a particular polypeptide. (Crick, 1963)

coding sequence: That portion of a eukaryotic gene separated by intervening sequences, which actually codes for the protein.

coding strand: The DNA strand with the same sequence as the transcribed mRNA (given U in RNA and T in DNA). *Compare* ANTICODING STRAND. According to recent terminology, coding strand is called *antisense strand*.

codominance: The situation whereby both members of a pair of alleles contribute to the phenotype.

codominant alleles: Both alleles contribute to the phenotype; neither is dominant over the other.

codon: A triplet of nucleotides that code for a single amino acid. Also known as *code word*. (Crick, 1963)

TABLE 1. The Genetic Code Dictionary: mRNA Codons

First letter (5′)	Second letter				Third letter (3′)
	U	C	A	G	
U	PHE PHE LEU LEU	SER SER SER SER	TYR TYR ter ter	CYS CYS ter TRY	U C A G
C	LEU LEU LEU LEU	PRO PRO PRO PRO	HIS HIS GLU GLU	ARG ARG ARG ARG	U C A G
A	ILE ILE ILE MET	THR THR THR THR	ASP ASP LYS LYS	SER SER ARG ARG	U C A G
G	VAL VAL VAL VAL	ALA ALA ALA ALA	ASP ASP GLU GLU	GLY GLY GLY GLY	U C A G

At initiation positions, UUG, AUG, and GUG are recognized by $tRNA_f^{met}$. At internal positions, UUG is recognized by $tRNA^{leu}$, AUG is recognized by $tRNA^{met}$, and GUG is recognized by $tRNA^{val}$. Termination codons: UAA, ocher; UAG, amber; UGA, opal.

codon bias: The idea that for amino acids with several codons, one or a few are preferred and are used disproportionately. They would correspond with tRNAs that are abundant. Also known as *codon preference*. (Air et al., 1976)

codon-anticodon relationship: This is alignment on the ribosome of an anticodon of an amino-acid-charged-tRNA molecule against its complementary codon on mRNA.

coefficient of inbreeding: The probability that offspring will be homozygous for a given trait if neither the parents nor the grandparents are homozygous for that trait and one of the grandparents is a carrier for that gene.

coefficient of integration: The relationship between the frequency of transfer of a donor marker to recipients and the frequency to its transmission from recipients to recombinants following F-X HFR conjugal mating. (Wright, 1929)

coefficient of migration: *See* MIGRATION COEFFICIENT.

coefficient of relationship (r): The proportion of alleles held in common by two related individuals.

coefficient of replacement: *See* MIGRATION COEFFICIENT.

coefficient of variation: The standard deviation expressed as a percentage of the mean.

cog region: Any group of similar sequences of nucleotides in DNA molecules that may specifically be recognized by endonucleases or other enzymes. (Angel, Austin, and Catcheside, 1970)

cognate tRNAs: tRNAs that are recognized by a particular aminoacyl-tRNA synthetase.

cohesive end: *See* STICKY END. (Ris and Chandler, 1963)

coincidence: This is the chance of a simultaneous occurrence of two events.

coincidence coefficient: The observed frequency of double crossovers, divided by their calculated or expected frequency. Expressed as a frequency, a measure of interference. In positive interference the coincidence is greater than 1. (Muller, 1916)

coincidental evolution: *See* CONCERTED EVOLUTION. (Hood, Campbell, and Elgin, 1975)

cointegrate plasmid: A plasmid carrying both genes for antibiotic resistance and genes for transfer.

cointegrate structure: A fusion of two elements. An intermediate structure in transposition.

colcemid: A synthetic equivalent of colchicine.

colchicine: A poisonous alkaloid, $C_{22}H_{25}NO_6$, derived from the autumn crocus, *Colchicum autumnale*. Used as an agent for interrupting mitosis in plant cells and in minute concentrations for the treatment and prevention of gout.

colchiploidy: colchicine-induced polyploidy.

colicin E1: A protein that kills *E. coli.*

colinear: Chromosomes containing the same sequence of genes.

colinearity: This term refers to the fact that a gene and the polypeptide for which it codes are related in a direct linear fashion, with the 3′ end of the template strand of the gene corresponding to the aminoterminus of the polypeptide. (Crick, 1963)

colony: A visible clone of cells.

colony hybridization: A technique for using *in situ* hybridization to identify bacteria carrying chimeric vectors whose inserted DNA is homologous with some particular sequence.

colony-forming efficiency: The percentage of cells plated (seeded, inoculated) that form a colony.

color inhibitor: One of the number of dominant genes that preclude color formation.

combination breeding: Combination desired characteristics of both the parents in one variety: no new characters or new intensity of given trait can be created in the resultant offsprings or recombinants. "Overcoming limiting factors" or "resistance breeding" through backcrossing is an example.

combining ability, general: The average or overall performance of a genetic strain in a series of crosses. (Griffing, 1956)

combining ability, specific: The performance of specific combinations of genetic strains in crosses in relation to the average performance of all combinations. (Griffing, 1956)

commaless code: Said of a genetic code in which successive codons are contiguous and not separated by noncoding bases or groups of bases.

commensalism: An association in which one symbiont benefits and the other is neither harmed nor benefited.

community: The group of organisms found in a particular place.

comparative mapping: Mapping a common set of probes onto linkage maps of different species. This process helps in establishing syntany of markers and unraveling the relationship among related genera. Comparative maps have been established for cereals such as wheat, maize, rice, sorghum, and among some dicots such as tomatoes, potatoes, and peppers.

compensating trisomic: An individual with an extra chromosome (2n + 1) in which a missing standard chromosome is compensated by two novel chromosomes, carrying two different arms of the missing chromosome. (Blakeslee, 1927; Khush and Rick, 1967)

competence factor: A surface protein that binds extracellular DNA and enables the cell to be transformed. (Charpak and Dedonder, 1965)

competitive exclusion: The principle that no two species can coexist in the same place if their ecological requirements are identical. This is also known as *Gause's Law.*

complementary bases: Refers to nucleotides or nucleotide sequences that are able to base pair. Adenine (A) and thymine (T) are complementary, and guanine (G) and cytosine (C) are complementary. Thus, the sequence 5′-ATGC-3′ is complementary to 5′-GCAT-3′.

complementary chiasmata: Chiasmata formed between four chromatids in such a way that the first chiasma occurs between two of the chromatids and the second between their homologs.

complementary DNA (cDNA): DNA produced on an RNA template.

complementary genes: Genes that interact to produce an effect distinct from the effects of an individual gene. Complementary genes yield the same mutant phenotype when present separately, but when present together they interact to produce a wild-type phenotype. Two heterogygotes in F_2 progeny produce ratio 9 wild-type:7 mutant.

complementary RNA: Synthetic RNA produced by transcription from a specific DNA single-stranded template.

complementary sequence: A nucleotide sequence that base pairs with another nucleotide sequence.

complementation: The ability of linearly adjacent segments of DNA to supplement each other in phenotypic effect. Complementary genes, when present together, interact to produce a different expression of a trait.

complementation assay *in vitro*: This assay consists of identifying a component of a wild-type cell that can confer activity on an extract prepared from a mutant cell. The assay identifies the component rendered inactive by the mutation.

complementation group: Cistron (determined by the *cis*-trans complementation test).

complementation map: A map developed from the complementation relationships between alleles, normally in a small segment of the chromosome.

complementation matrix: A tabular representation of complementation tests involving a number of phenotypically similar mutants.

complementation test: *See* CIS-TRANS COMPLEMENTATION TEST.

complete digest: The treatment of a DNA preparation with a restriction endonuclease for sufficient time that all of the potential target sites within that DNA have been cleaved. *Compare* PARTIAL DIGEST.

complete dominance: Resemblance of F_1 with one of its parents.

complete linkage: A condition where no crossovers are observed in a test cross or F_2 progeny. Closely linked genes may show complete linkage. In some species, crossing-over is absent in one of the sexes. This is also an example of complete linkage.

complete penetrance: Said of a dominant gene whose presence in an organism always produces an effect on that organism, or of a recessive gene whose presence in the homozygous state always produces an effect. *See* PENETRANCE and INCOMPLETE PENETRANCE.

completely randomized design: An experimental design that is used when the experimental material is limited and homogeneous.

complex character: One whose difference from an alternative is not transmitted as a single unit of heredity.

complex gene: A gene or its protein product that undergoes various rearrangements, cleavages, or modifications before a functional product is obtained.

complex transcription unit: When a primary transcript yields more than one mRNA encoding different proteins.

complexity of DNA: It is the total length of different sequences of DNA present in a given preparation.

complon: A subunit of a complementation map of a cistron.

component of fitness: A particular aspect in the life cycle of an organism upon which natural selection acts.

composite transposon: (1) A transposon constructed of two IS elements flanking a control region that frequently contains host genes; (2) DNA sequence flanked by insertion sequences that is capable of moving as a unit, thus facilitating its transposition. *See* INSERTION ELEMENTS and INSERTION SEQUENCES (IS ELEMENTS).

composites: A mixture of genotypes from several sources, maintained by normal pollination. *See* SYNTHETICS.

compound cross: Combining desirable genes from more than two inbred lines; combining characters of many varieties into one plant.

compound gene: The genes in which coding sequences are separated by noncoding sequences. Also known as *split genes*.

concept of dominance: The concept that dominance is not a universal property of a gene. It is relative property of a gene to produce a particular phenotype in a particular genetic and environmental background.

concerted evolution: The ability of two related genes to evolve together as though constituting a single locus. (Zimmer et al., 1980; Dover and Coen, 1981)

condensation of chromosomes: The spiralization and charging of a chromosome with nucleic acid. Also known as *contraction of chromosomes*.

condensation reaction: A reaction in which a covalent bond is formed with loss of a water molecules, as in the addition of an amino acid to a polypeptide chain.

conditional mutation: Any mutation whose manifestation is dependent on one or more parameters, such as temperature, age, or nutrition. (Edgar and Lielausis, 1964)

conditional-lethal mutant: A mutant that is lethal under one condition but not lethal under another condition.

conditioned dominance: Dominance affected by the presence of other genes or dominance affected by environmental variation.

confidence limits: A statistical term for a pair of numbers that predict the range of values within which a particular parameter lies.

configuration: An association of two or more chromosomes at meiosis segregating independently of other associations at anaphase.

consensus sequence: An average sequence of which each nucleotide is the most frequent at that position in a set of examples; used for RNA splice sites and other sites.

conservative recombination: The recombination that involves breakage and reunion of preexisting strands of DNA without any synthesis of new stretches of DNA.

conservative replication: A postulated mode of DNA replication in which an intact double helix acts as a template for a new double helix; this type of replication is known to be incorrect.

conserved sequence: A sequence found in many different DNA or RNA samples (e.g., promoters) that is invariant in the sample.

constitutive enzyme: An enzyme synthesized at a regular rate regardless of the environment. *See* INDUCIBLE ENZYME.

constitutive gene: A gene whose expression is not regulated. Its product is continuously synthesized by the cell—whether or not substrate is present in the cell.

constitutive heterochromatin: Permanently condensed and inactive chromatin. Present in both the homologues at identical positions. Composed of highly repetitive DNA sequences. *See* SATELLITE DNA.

constitutive mutant: A mutant whose transcription is no longer under regulatory control.

constitutive mutation: A mutation that causes the genes that usually are regulated to be expressed without regulation.

"contact" hypothesis: The theory that chromosomes occasionally undergo illegitimate crossing-over between nonhomologous sections, resulting in an interchange of blocks of genes. (Dobzhansky, 1946)

contact inhibition: The phenomenon whereby contact between cells in culture prevents their further movement. (Abercrombie and Haeysman, 1953)

containment: The precautions used to prevent the replication of the products of recombinant DNA technology outside the laboratory. *Biological containment* restricts the use of hosts and cloning vectors thus rendering recombinant products less able to survive and multiply outside the laboratory environment. *Physical containment* deals with use of special laboratories that prevent escape of products of recombinant DNA technology.

contig: A sequence of overlapping contiguous chromosome segments.

contingency table: A table of frequencies showing two classifications simultaneously that is used in testing their independence.

continuity, Yates' correction for: A correction applied in the calculation of normal deviates of X^2s to allow for the discrepancy arising by the observation being discontinuous while tables of the normal deviate and X^2 are calculated on the supposition of continuity in the variate.

continuous culture: The culture of microorganism in liquid medium under controlled conditions, with additions to and removals from the medium over a length of time.

continuous replication: In DNA, uninterrupted replication in the 5′ to 3′ direction using a 3′ to 5′ template.

continuous trait: Quantitative trait; variation that cannot be represented by discrete classes and that requires measurement data. Multigenic or polygenic inheritance represents this type of variation.

continuous variation: Variation measured on a continuum rather than in discrete units or categories (e.g., height in human beings).

contractile ring: A ring of actin filaments that forms around the equator at the end of mitosis and is responsible for pinching the daughters cells apart. (Lewis, 1954)

contrasting genetic characters: Genetic characters with marked phenotypic differences.

controlling elements: These elements of maize are transposable units originally identified solely by their genetic properties. They may be autonomous (able to transpose independently) or nonautonomous (able to transpose only in the presence of an autonomous element). (McClintock, 1951)

conventional pseudogene: A DNA sequence that resembles a gene but has become inactive due to the accumulation of mutations.

convergence: A superficial resemblance between organs or organisms that has evolved under the influence of similar adaptive zones. (Winkler, 1930)

convergence-divergence selection: Selection of promising genotypes in a bulk population grown at diverse locations followed by massing of all selections and allowing mating among them in a pollination field. The bulk seeds from this pollination field constitute the base for the next cycle. This obviously increases ecological tolerance of the population.

convergent improvement: Double backcross method for simultaneous improvement of the two parents involved in a single cross for improving per se performance of lines without interfering with their combining ability.

coordinate gene regulation: The common control of a group of genes.

coordinate repression: Correlated regulation of the structural genes in an operon by a molecule that interacts with the operator sequence. (Ames and Garry, 1959)

copper fist: Configuration of a DNA-binding protein that resembles a fist closed around a penny. In this case, the penny is copper ions; the knuckles of the fist of the yeast ACE1 protein interact with the promoter of the metallothionein gene, enhancing its transcription.

copy DNA: *See* COMPLEMENTARY DNA.

copy number: (1) The number of molecules of a plasmid contained in a single cell; (2) the number of copies of a gene, transposon, or repetitive element in a genome; (3) the number of nucleic acid sequences that a probe contains.

copy-choice hypothesis: An incorrect hypothesis which stated that recombination resulted from the switching of the DNA-replicating enzyme from one homolog to the other. (Lederberg, 1955)

cordycepin: It is 3′ deoxyadenosine, an inhibitor of polyadenylation of RNA.

core DNA: The 146 bp of DNA contained in a core particle.

core enzyme: The form of a multisubunit enzyme from which a subunit has dissociated. *See* HOLOENZYME.

core octamer: The structure comprising two subunits each of histones H2A, H2B, H3, and H4, that forms the central component around which DNA is wound to form a nucleosome.

core particle: A digestion product of the nucleosome that retains the histone octamer and has 146 bp of DNA; its structure appears similar to that of the nucleosome itself.

core promoter: A eukaryotic promoter very similar in function to a bacterial promoter, which is located close to the coding region of a gene.

corepressor: A small molecule that must bind to a repressor protein before the latter is able to attach to its operator site. Effectors of repressible operons are often called corepressors. (Jacob and Monod, 1961a)

correction: The production (possibly by excision and repair) of a properly paired nucleotide pair from a sequence of hybrid DNA that contains an illegitimate pair.

correlated response: Change of the phenotype occurring as an incidental consequence of selection for a seemingly independent character, such as sterility resulting from selection for high bristle number.

correlation coefficient: A statistic that gives a measure of how closely two variables are related. Measure of the agreement or lack of agreement of two sets of observation.

cos site: The sequence with the λ DNA molecule that forms single-stranded overhangs at the ends of the linear version of the genome.

cosduction: Transduction of foreign DNA by use of cosplasmid (a plasmid onto which the cohesive ends form). The transducing phage particles are called *cosducing particles* and the recipients *cosductants*. (Flock, 1983)

cosegregation: In *Chlamydomonas,* parallel behavior of different chloroplast markers in a cross, due to their close linkage on cpDNA.

cosmid: A hybrid plasmid that contains cos sites at each end. Cos sites are recognized during head filling of lambda phages. Cosmids are useful for cloning large segments (up to 45 kb) of foreign DNA. (Collins and Hohn, 1978)

cost of selection: The proportion of adaptively inferior genotypes that are produced in a population in the course of producing adaptively superior genotypes. (Haldane, 1957)

C_0t: The product of DNA concentration and time of incubation in a reassociation reaction. (Britten and Kohne, 1968)

C_0t curve: A plot of percentage reassociation of denatured DNA as function of C_0t units.

C_0t plot: A graph showing the extent of reannealing of previously denatured DNA molecules as function of C_0t value. C_0t plots indicate the relative amounts of unique sequence, moderately repetitive and highly repetitive DNA in a genome, as giving some idea of the total complexity of sequences.

C_0t values ($C_0t_{½}$): The product of C_0, the original concentration of denatured, single-stranded DNA and t, time in seconds, giving a useful index of renaturation. $C_0t_{½}$ values are the midpoint values on C_0t curves (C_0t values plotted against concentration of remaining single-stranded DNA), which estimate the length of unique DNA in a sample.

cotransduction: Simultaneous transduction of two or more bacterial genes by a single bacteriophage particle. The cotransduction frequency is inversely related to the distance between the genes on the chromosome. Cotransduction frequency = $(1 - d/l)^3$ where d = distance between markers, and l = length of the transduced gene (exogenote).

countertranscript: Any RNA species that is initiated from a single start within the leader region and transcribed from the opposite strand, one terminating very close to one mRNA start, the other further upstream. One or both of these countertranscripts serves as negative control elements. (Kumar and Novick, 1985)

coupling linkage: The tendency of two mutant characters, inherited by F_1 from one parent, to stay together in F_2. Also known as *cis-arrangement*. *See* REPULSION. (Bateson, 1905)

covalently closed circle (CCC): A double-stranded DNA molecule with no free ends. The two strands are interlinked and will remain together even after denaturation. In its native form, a CCC will adopt a supercoiled configuration.

covariance: A statistical value measuring the simultaneous deviations of x and y variables from their means.

CpG islands: A region of genomic DNA that contains a very high proportion of CpG residues, and a dinucleotide that is often located at the 5′ end of genes.

critical chi-square: A chi-square value for a given degree of freedom and probability level to which an experimental chi-square is to be compared.

critical difference: The values indicating least significant difference at values greater than which all the differences are significant.

cron: The time unit for evolutionary processes (1 cron = 10^6 years; 1 kilocron = 10^9 years; 1 millicron = 10^3 years). (Huxley, 1957)

cross: A bringing together of material from different genotypes to achieve recombination.

crossbreed: Fertilization between separate individuals.

crossbreeding: Mating between members of different races or species.

cross-fertilization: Fertilization of a female gamete with a male gamete derived from a different individual.

cross-hybridization: The annealing of two nucleic acid molecules that are not perfectly complementary.

crossing: Artificial pollination between genetically dissimilar plants.

crossing-over: A process inferred genetically from the recombination of linked genes in the progeny of heterozygotes. A process evidenced cytologically from the formation of chiasmata between homologous chromosomes during diplotene stage. Exchange of corresponding segments between chromatids of homologous chromosomes, by breakage and reunion, following pairing. (Morgan and Cattell, 1912)

crossover: Said of a chromatid or gamete resulting from crossing-over.

crossover fixation: A possible consequence of unequal crossing-over that allows a mutation in one member of a tandem cluster to spread through the whole cluster (or to be eliminated).

crossover suppression: The apparent lack of crossing-over within an inversion loop in heterozygotes, which is due to mortality of zygotes carrying defective crossover chromosomes rather than actual suppression.

crossover unit: The percent of recombination of linked genes. One percent of recombination equals one crossover unit in a linkage map.

crossover value: The percentage of crossing-over in a hybrid population; a term used mostly in determining linkage percentage, particularly in chromosome mapping.

cross-pollination: The transfer of pollen from an anther to the stigma of a flower of a different plant.

crown gall: A tumor formed, usually, on the stems of broad-leaved plants when infected with *Agrobacterium tumefaciens* containing a Ti plasmid. The bacterium is only necessary for the initiation of the tumor. The genome of the affected plant cells contains several copies of a segment of the Ti plasmid (the T-DNA). Crown galls can be of two types, octopine or nopaline, depending on the type of Ti plasmid that initiated the tumor. Whole plants can be regenerated from crown-gall tissue and some of these will contain the T-DNA.

cruciform structure: The cross-shaped structure that can arise by intramolecule pairing within a double-stranded DNA molecule which contains an inverted repeat.

cryopreservation: *See* FREEZE PRESERVATION.

cryoprotectant: An agent that will protect cells or tissues from the deleterious effects of freezing. Such agents include glycerol and dimethylsulfoxide.

cryptic polymorphism: A type of polymorphism that is controlled by recessive genes.

cryptic satellite: A satellite DNA sequence not identified as such by a separate peak on a density gradient; that is, it remains present in main-band DNA.

cryptic species: The species not distinguishable by external morphology.

cryptic variation: Differences among individuals in terms of some biochemical properties (e.g., denaturation) that are not revealed at the morphological level or even electrophoretically.

cryptomere: (1) A gene which by itself has no visible effect but whose existence can be demonstrated by means of suitable crosses, i.e., a complementary factor; (2) a recessive factor.

cryptomorphic gene: The gene has a cryptic structure in that the ultimate active product or products are carried within the precursorial molecule. The active protein is released after enzymatic breakdown of the precursor followed by processing.

CsCl—caesium (cesium) chloride: A salt that forms very dense solutions in water or aqueous buffers and is used in isopycnic centrifugation to separate DNA molecules of different densities. When spun at a high speed in an ultracentrifuge, Cs^+ atoms distribute themselves (ca. 1.5 to 1.8 g cm-^{2}) for the separation of DNA molecules.

cultivar: Synonymous with variety; the international equivalent of variety.

cultivar identification system: This system based on STMS (sequence-tagged microsatellite sites) loci analysis with fluorescent primers and the "Genescan" software allows unequivocal identification of cultivars, paternity testing, as well as identification of duplicates using an electronic database.

cultural evolution: A change in the relationship between the organism and the environment. The organism adapts the environment according to its need. Its rate is fast compared to biological evolution. "Word" is the unit of cultural heredity. Forces of cultural evolution are conscious selection, foresightedness, planning, etc.

cumulative effect: The action of two alleles of a gene giving a more pronounced effect than one in heterozygous conditions. Probably one allele supplies insufficient enzyme for the reaction conditioned by two. The hybrids are distinguishable from parents. The effect is sometimes erroneously referred to as incomplete dominance.

cumulative genes: *See* POLYGENES. (Nilsson-Ehle, 1911)

C-value paradox: The estimated number of genes in eukaryotes is much less than the amount of DNA present, e.g., haploid human genome has 2.8×10^7 bp and should contain 3×10^6 genes, but the number estimated is 50,000. This anomalous situation is referred to as C-value paradox. (Cavalier-Smith, 1978)

cybrid: The viable cell resulting from the fusion of a cytoplant with a whole cell, thus creating a cytoplasmic hybrid. *See* VYBRID. (Bunn, Wallace, and Eisenstadt, 1974)

cyclic AMP: A modified version of adenosine monophosphate, which through an intramolecular phosphodiester bonds involving $3'$ and $5'$ positions of sugar, assumes a circular shape.

cyclobutyl dimer: A dimeric structure formed between two adjacent pyrimidine bases in a polynucleotide, resulting from ultraviolet irradiation.

cytidine: The ribonucleoside containing the pyrimidine cytosine.

cytidylic acid: The ribonucleotide containing the pyrimidine cytosine.

cytochemistry: The analysis, often by determination of staining properties, of the chemical composition of cell components.

cytochromes: A class of proteins, found in mitochondrial membranes, whose main function is oxidative phosphorylation of ADP to ATP.

cytogenetic map: A map showing the actual position of the gene within the physical chromosome, distinct from a chromosome map, which merely shows the position of the genes relative to each other.

cytogenetics: The correlated study of the transmission and recombination of genes and the behavior of the chromosomes during mitosis and meiosis. It is a field of investigation related to assigning genes to specific regions of the chromosomes. (Sutton, 1903)

cytogenic male sterility: Whenever restorer line becomes available for cytoplasmic male sterile line, sterility is called cytogenic, cytoplasmic-genic, or cytoplasmic-genetic male sterility.

cytohet: A genetic condition in which a zygote contains in its cytoplasm genetically different mitochondria contributed by two parents; thus, the individual is cytoplasmically heterozygous. (Sagar, 1973)

cytokinesis: The division of the cytoplasm during cell division in mitotic cells.

cytokinesis I: The division of the cytoplasm during meiosis I division giving rise to two haploid cells.

cytokinesis II: The division of the cytoplasm during meiosis II division giving rise to four haploid cells.

cytokinins: A class of growth hormones that cause cell division, cell differentiation, shoot differentiation, etc. Some of the cytokinins commonly used in tissue culture are kinetin, benzylamino-purine (BA), 2-isopentenyladenine (2-ip), and zeatin.

cytological map: The assignment of genes to particular chromosomes by observation of chromosomal morphology.

cytology: The study of the structure and function of cells and their organelles.

cytophotometry: Photometric analysis of different macromolecules (DNA, RNA, proteins, and polysaccharides) in cells or tissue sections.

cytoplasm: The protoplasm of a cell other than the nucleus. (Strasburger, 1882)

cytoplasmic: Pertaining to or centered in the cytoplasm.

cytoplasmic DNA: The DNA that is found in the cytoplasm in chloroplasts and mitochondria.

cytoplasmic inheritance: Inheritance of characters whose governing genes are located in the cytoplasmic organelles (mitochondria, chloroplasts) rather than in nucleus of eukaryotes. Also known as *extra-nuclear inheritance*. In prokaryotes, genes contained in plasmids show similar inheritance but is called *extrachromosomal inheritance*.

cytoplasmic male sterility (CMS): Maternally inherited inability in higher plants to produce fertile pollen. Formation of chimeric genes in mitochondria is the cause of CMS. (Welch and Grimball, 1947)

cytoplasmic segregation and recombination (CSAR): A process suggested to explain assortment and recombination of organelle-based genes.

cytoplast: The intact cytoplasm remaining following the enucleation of a cell.

cytoplastic inheritance: Transmission of hereditary characters through the cytoplasm as distinct from transmission by genes carried by chromosomes. Detected by differing contribution of male and female parents in reciprocal crosses. *See* CYTOPLASMIC INHERITANCE.

cytosine: 2-oxy, 6-aminopyrimidine, one of the pyrimidine bases found in DNA and RNA.

cytoskeleton: Networks of fibers in the cytoplasm of the eukaryotic cell. (Goldman et al., 1975)

cytosol: The general volume of cytoplasm in which organelles (such as the mitochondria) are located.

dark repair: The repair of DNA damage by DNA repairing enzymes that do not require light for their activation.

Darwinian evolution: A theory of evolution proposed by Charles Darwin that explains evolution in terms of natural selection. *See* DARWINISM.

Darwinian fitness: The fitness of a given genotype in a given environment is measured by its relative contribution to the ancestry of future generations—that is, by the change in the frequency of this genotype from one generation of parents to the next generation of parents. It depends on both fertility and survival.

Darwinism: The hypothesis of Darwin according to which evolution of new species is by the accumulation of small changes. "Natural selection" and the "survival of the fittest" play an important role in this.

daughter cells: Young cells derived by the division of an older cell (mother cell).

daughter chromosomes: The two chromosomes produced by the replication of a single parental chromatid previously associated with one centromere at metaphase I or metaphase II.

de novo: From Latin, meaning "arising anew, afresh, once more."

DEAE-cellulose: The complete name is *diethylaminoethylcellulose*. It is the ion exchange material that binds anions; used in chromatography especially for proteins.

decay of genetic variability: The reduction of heterozygosity because of genetic drift leading to loss and fixation of alleles at various loci.

decoding: *See* RECODING. (Gesterland, Weiss, and Atkins, 1992).

dedifferentiation: The resumptions of meristematic activity by more or less mature cells through a reversal of the process of cell or tissue differentiation. Cell division leading to the formation of small, microvacuolate, isodiametric cells with prominent nuclei, which are capable of organized development, as when adventitious buds or roots develop on mature tissue.

deficiency: *See* DELETION.

deficiency loop: The loop formed by the nonsynapsing portion of an unaltered chromosome, caused by a deficiency in the homologous chromosome.

definitive nucleus: A nucleus formed in the embryo sac by the fusion of the two polar nuclei. The definitive nucleus later fuses with a male nucleus to form the triploid endosperm nucleus.

degenerate code: The genetic code that is characterized by the fact that two or more codons may designate the same amino acid.

degree of dominance: A ratio of additive genetic variance to total phenotypic variance. If this ratio equals 1, the trait shows complete dominance, if the ratio is greater than 1, the trait shows overdominance, and if the ratio is less than 1, the trait shows incomplete dominance.

degrees of freedom: An estimate of the number of independent categories in a particular statistical test or experiment.

delayed inheritance: The inheritance whereby each generation manifests, with regard to particular characters, the genotype of the female parent.

deletion: The loss of a part of a chromosome, usually involving one or more genes (rarely a portion of one gene). (Painter and Muller, 1929)

deletion analysis: The identification of regulatory sequences for a gene by determining the effect on gene expression of specific deletions in the upstream region.

deletion heterozygote: An individual heterozygous for the absence of an internal segment of one of its chromosomes.

deletion loop: The loop formed by the nonsynapsing portion of an unaltered chromosome, caused by a deletion in the homologous chromosome.

deletion mapping: Mapping mutation by use of overlapping deletion mutants to determine whether or not a mutation includes the site of a mutant gene.

deletion series: A set of clones or derived from the same initial recombinant, but in which the insert lack sequences at one of its ends because of treatment with an exonuclease.

demethylation model: A model that explains the role of undermethylation of cytosine in gene regulation and differentiation in eukaryotes. In undifferentiated cells, DNA is fully or uniformly methylated at all sites that ever will be methylated. During development, sequence-specific proteins inhibit methylation during DNA replication, leading to methylation patterns specific for each tissue. Once demethylation events occur, the differentiated methylation patterns are clonally inherited as a result of the maintenance methylation system.

demic diffusion: The process of colonization of an area by a population. This term is used in contrast with cultural diffusion, which involves movement of ideas rather than petition.

demic selection: Disproportionate growth of a subset of a population or species. Demic selection has an effect on the general genetic composition of the population if the subsets have different gene frequencies (as is usually the case). Demic selection is a type of intergroup selection that does not necessarily involve direct competition.

demography: The study of populations and how they survive, die, reproduce, and grow.

denaturation: Breakdown by chemical or physical means of the noncovalent interactions that are responsible for the secondary and higher levels of structure of proteins and nucleic acids.

denaturation map: A map of a stretch of DNA showing the locations of local denaturation loops, which correspond to regions of high AT content.

dendrogram: An evolutionary tree diagram that orders objects (e.g., individuals, genes) on the basis of similarity.

Denhardt's solution: A solution used in DNA hybridization procedures such as the Southern biotechnique. It is used to coat a nitrocellulose filter, thus blotting to inhibit nonspecific binding of the probe.

density gradient centrifugation: A method of separating molecular entities by their differential sedimentation in a centrifugal gradient.

density-dependent inhibition of growth: Mitotic inhibition correlated with increased cell density.

deoxyadenosine: The deoxyribonucleoside containing the purine adenine.

deoxyadenylic acid: The deoxyribonucleotide containing the pyrimidine adenine.

deoxycytidine: The deoxyribonucleoside containing the pyrimidine cytosine.

deoxycytidylic acid: The deoxyribonucleotide containing the pyrimidine cytosine.

deoxyguanosine: The deoxyribonucleoside containing the purine guanine.

deoxyguanylic acid: The deoxyribonucleotide containing the purine guanine.

deoxyribonuclease: An enzyme that breaks a DNA polynucleotide by cleaving phosphodiester bonds.

deoxyribonucleic acid (DNA): A usually double-stranded, helically coiled, nucleic acid molecule composed of deoxyribose-phosphate "backbones" connected by paired bases attached to the deoxyribose sugar; the genetic material of all living organisms and many viruses.

deoxyribonucleoside: A portion of a DNA molecule composed of one deoxyribose molecule plus either a purine or a pyrimidine.

deoxyribonucleotide: A portion of a DNA molecule composed of one deoxyribose phosphate bonded to either a purine or a pyrimidine.

deoxyribose: The 5-carbon sugar of DNA.

depurination: Removal of adenine and guanine from DNA or RNA thus forming apurinic acid from purines. This activity breaks the sugar-phosphate backbone in the nucleic acids.

derepression: (1) Initiation of or increase in transcriptional activity of a gene sequence resulting from the removal of a repressor molecule from the sequence or a neighboring control sequence. (2) The mechanism by which repression is alleviated, as when a repressing metabolite is removed, resulting in the increased level of a protein or enzyme.

derivative hybrid: A hybrid arising from a cross between two hybrids.

desmutagen: An agent that suppresses mutagenic activity of a mutagen by directly inactivating the mutagen by destroying the oxygen radicals produced. (Kada et al., 1985)

desynapsis: The falling apart of chromosomes during diplotene or diakinesis that paired during zygotene and remained paired at pachytene. (Li, Pao, and Lich, 1945)

detection of linkage: Statistical analysis in which the total X^2 is partitioned into various components to determine the presence of linkage in F_2 or test-cross progeny.

determination: The process by which a cell line becomes genetically fixed to differentiate into a specific type of tissue or organ. (Roux, 1905)

deterministic events: Events that have no random or probabilistic aspects but proceed in a fixed, predictable fashion.

detrimental allele: Any gene form that imparts a disadvantage to the individual carrying it and decreases the carrier's chances of reproduction or survival compared with that of an individual not possessing the allele. (Muller, 1934)

development: The process of orderly change that an individual goes through in the formation of structure. More specifically, development is the translation of unidimensional information of the gene in the chromosomes into a two-dimensional blastoderm, which then is transformed into a three-dimensional adult consisting of a large number of tissues and organs.

developmental constraint: The ontogenetic rules that determine the course of evolution; the degree of intrinsic control over evolutionary direction (in contrast to extrinsic natural selection).

developmental correlation: The pleiotropy resulting from the influence of the same gene upon character differences that are necessarily affected by the same developmental or metabolic pathway during ontogeny.

developmental genetics: A study of the operation of genes during development.

developmental homeostasis: The capacity of the developmental pathway to produce a normal phenotype in spite of developmental or environmental disturbances. (Lerner, 1954)

developmental noise: The influence of random molecular processes on development.

developmental trial: Mini-kit trials that are conducted by a large number of farmers under their own socioeconomic conditions to determine whether they can grow a new variety. It is a sort of farmer's participation in varietal evaluation before the new varieties are finally released by the breeders for cultivation.

deviation: A departure from the expected or from the expected value. (de Beer, 1930)

DeVriesism: The hypothesis that evolution in general and speciation in particular are the results of drastic mutations (saltations). *See* MUTATION THEORY OF EVOLUTION. (DeVries, 1901)

dG:dC tailing: A method used for molecular cloning of random fragments of DNA by adding complementary homopolymer tails to the insert and the vector DNA. Oligo dG and oligo dC tails are added by a terminal transferase.

diakinesis: A substage of prophase I of meiosis I during which chromosomes have separated except at the ends. The homologous chromosomes are present as bivalents. (Haecker, 1897)

diallele crossing: The system where each of a member of males is mated to each of a number of females. Also known as *diallele mating*. *Complete diallele* results into n^2 progenies and *half diallele* into $n(n - 1)/2$ progenies.

diallele selective mating: Enlarging the initial hybrid base (otherwise nonexistent in self-fertilizing plants) by developing three or four composite hybrid pools (populations) from multiple crossing in diallele fashion.

dicentric chromosome: A chromosome or chromatid with two centromeres. (Darlington, 1937)

dichotomous regulators: Dual-function regulators that activate transcription in one circumstance and repress in another. (Saur et al., 1995)

dichotomous transcription elongation: Elongation of transcription proceeds in alternating laps of monotonous and inchworm-like movement with the flexible RNA polymerase configuration being subject to direct sequence control. (Nudler, Goldfarb, and Kashlev, 1994)

dideoxynucleotide: A modified nucleotide that lacks the 3′ hydroxyl group and so prevents further chain elongation when incorporated into a growing polynucleotide.

dideoxynucleotide method of DNA sequencing: A method for DNA sequencing that is based on the ability of a DNA polymerase to extend a primer, annealed to the template that is to be sequenced, until a chain-terminating nucleotide ddNTP is incorporated. The resulting series of unique fragments are separated by polyacrylamide gel electrophoresis. The basic procedure involves the following: (a) hybridization of an oligonucleotide primer to a suitable single- or denatured double-stranded DNA template; (b) extension of the primer with DNA polymerase in four separate reaction mixtures, each containing one α-labeled dNTP, a mixture of unlabeled dNTPs, and one chain-terminator ddNTP; (c) resolution of the four sets of reaction products on a high-resolution polyacrlamide/urea gel; and (d) production of an autoradiographic image of the gel that can be examined to deduce the DNA sequence. (Sanger, Nicklen, and Coulson, 1977)

differential affinity: The failure of two chromosomes to pair at meiosis in the presence of a third, although they pair in its absence. (Darlington, 1928)

differential processing: This refers to production of more than one functional mRNAs from one type of complex transcription unit through differential processing.

differential reproduction: Reproduction in which different types do not contribute to the next generation in proportion to their numbers.

differential screening: A method of screening cDNA libraries whereby two probes, differing in one or just a few sequences, are used as probes in two parallel *in situ* hybridizations using duplicate lifts of the library.

differential segments: Portions of chromosomes that do not pair in meiosis.

differential splicing: A form of splicing wherein the combination of introns that are removed varies between cell types, allowing a single gene to produce transcripts which differ in primary structure and therefore in coding potential. *See* ALTERNATIVE RNA SPLICING.

differential staining: Chromosomes, when denatured and subsequently renatured and stained with some dye (such as quinacrine mustard or Giemsa), show some darkly stained bands along the chromosome. This technique is very useful for identifying the individual chromosomes.

differentiation: In biology, development of different types of cells and tissues from a single-celled zygote. The cells that have become specialized for a particular function are known as *differentiated cells*. These cells maintain, in culture, all or much of the specialized structure(s) or function(s) typical of the cell type *in vivo*.

digenic inheritance: Pattern of inheritance observed in a cross between two individuals identically heterozygous at two gene pairs. Such inheritance has a 9:3:3:1 ratio in F_2.

dihaploids: Diploid plants are obtained from colchiploiding haploids. In pollen or anther culture, dihaploid plants appear spontaneously (may be called *somadihaploid*), or they can be developed through colchicine treatment.

dihybrid: An individual that is heterozygous in two pairs of alleles.

dihybrid cross: A sexual cross in which the inheritance of two pairs of alleles is followed (tall, yellow × dwarf, green).

diiso-compensating trisomic: A compensating trisomic in which a missing chromosome is compensated for by two isochromosomes, one for each arm of the missing chromosome. (Kimber and Sears, 1968)

diisosomic: An individual that is missing one chromosome pair but has two homologous isochromosomes for the same arm of the missing pair. (Kimber and Sears, 1968)

diisotrisomic: An individual deficient in one chromosome but having a pair of isochromosomes for one arm of the missing chromosome. (Kimber and Sears, 1968)

dikaryon: A pair of haploid nuclei differing in character that can be genetically complementary to each other. (Maire, 1902)

dikaryosis: The condition of possessing two nuclei in each cell.

dimerization: The chemical union of two identical molecules.

dimonoecious: Having perfect (bisexual) flowers as well as pistillate and staminate flowers on one plant.

dimorphic gene: A gene whose protein product exists in two forms—inactive and active; active form is produced by cleaving a part of the polypeptide chain of the inactive protein.

dimorphism: A "polymorphism" involving only two forms of a biological property.

dinucleotide: Two nucleotides linked together through their phosphate groups. Two important dinucleotide are the coenzyme flavin adenine dinucleotide (FAD) and nicotinamide adenine dinucleotide (NAD).

dioecious: Individuals producing either sperm or egg but not both. In dioecious species, the sexes are separate. *Compare* MONOECIOUS.

diplo-chromosome: A chromosome that has divided twice instead of once since the preceding effective mitosis, with its centromere being undivided. (White, 1935)

diploid: The state of the cell in which all chromosomes, except sex chromosomes, are two in number and are structurally identical with those of the species from which the cultures were derived. (Strasburger, 1905)

diploidization: The process by which a polyploid species behaves like a diploid species.

diplonema: *See* DIPLOTENE.

diplontic selection: Selection of somatic mutants under the competition between mutated and surrounding nonmutated cells forming chimera. This is true for colchiploids as well.

diplophase: Part of the life cycle in which the zygotic chromosome number is found in reproductive cells other than gametes. (Belling, 1927)

diplosis: Establishment of the zygotic chromosome number, usually by syngamy and karyogamy. (Renner, 1916)

diplospory: A mode of apomixis in plants in which a diploid gametophyte is formed after mitotic or partly meiotic divisions of the spore-forming cells.

diplotene: A substage of prophase I of meiosis I during which chromosomes start separating and chiasmata are visible at the sites where crossing-over had occurred. (Winiwarter, 1900)

direct cross: A cross between two individuals.

direct repeats: Two or more identical nucleotide sequences present in a single polynucleotide.

directed alternate disjunction: Regular movement of alternate centromeres to the same pole of the spindle in nuclei heterozygous for one or more reciprocal translocations.

directed mutation: Selective mutation of a particular gene sequence or class of sequence. Such mutations may be achieved by the use of drugs that have a particular affinity for the sequence in question (e.g., active as compared to inactive gene sequences) or that tend to alter permanently the stage of differentiation of cells in culture.

directional dominance: Majority of dominant alleles having positive effects alienated in one direction and when having negative effects in the other, lead to genetic asymmetry in a population.

directional gene frequency: Greater frequency of dominant alleles than that of recessive alleles or vice versa in a population. This leads to asymmetry in the distribution of gene frequency.

directional selection: A type of selection that removes individuals from one end of a phenotypic distribution, thus causing a shift in the distribution.

disassortative mating: The mating of two individuals with dissimilar phenotypes.

disc gel electrophoresis: A technique in which a gel, usually of agarose or polyacrylamide, is cast in a vertical tube instead of a slab. The nucleic acid or protein bands are resolved as a series of discs in the cylindrical gel. The method is seldom used now mainly because of the difficulty of ensuring that each individual gel is run in an identical manner. However, disc gels, and their variants, still find application in preparative procedures.

discontinuous gene: A gene in which the biological information is divided between two or more exons, separated by introns.

discontinuous replication: In DNA the replication in short 5′ to 3′ segments using the 5′ to 3′ strand as a template while going backwards, away from the replication fork.

discontinuous trait: Variation in which discrete classes are easily recognized (e.g., tall vs. dwarf plants).

discontinuous variation: Variation that falls into discrete categories (e.g., the color of garden peas).

discordant: Members of a pair showing different, rather than similar, characteristics.

discrete generations: Generations that have no overlapping reproduction. All reproduction takes place between individuals in the same generation.

disease resistance: The ability to resist disease or the agent of disease (the vector) and so remain healthy. Resistance or tolerance to disease is a topic of intensive interest in plant tissue-culture work. Screening or selection at the cellular level is followed by plant regeneration and screening of these and their progeny.

disease Δ: Onset of disease requires interaction between a susceptible host, a virulent pathogen, and a favorable environment.

disjunct: A term used to denote a population whose geographical range does not contact that of the remainder of the species.

disjunct populations: Populations whose ranges do not overlap.

disjunction: The separation of homologous chromosomes during anaphase I of meiosis.

disomic: An individual with euploid chromosome complement. (Blakeslee, 1921)

disomic alien addition line: A line with an extra chromosome pair from a related species.

disomic haploid: An aneuhaploid that is disomic for one chromosome out of the total haploid complement ($x + 1$, $2x + 1$, $3x + 1$). It is derived from a tetrasomic.

disomic inheritance: Normal inheritance dependent upon two alleles (sometimes more) for a given gene, customarily found in diploid species with no extra chromosomes or chromosome sets (genomes).

dispersive replication: A postulated mode of DNA replication combining aspects of conservative and semiconservative replication; known to be incorrect.

disposable soma theory: The somatic cells of an organism are expendable. Aging results from destruction caused by molecules (including oxygen-free radicals) produced in the normal course of living. When these wear-and-tear processes outpower the repair systems, aging occurs.

disruptive selection: A type of selection that removes individuals from the center of a phenotypic distribution and thus causes the distribution to become bimodal. (Mather, 1948)

distal protein: With reference to the assembly of complex structures, e.g., ribosomes, these are proteins that bind late in the assembly process, and whose binding is dependent on the prior binding of proximal proteins. *See* PROXIMAL PROTEIN.

distinguishable hybrid: A hybrid in which intermediate inheritance is expressed (i.e., the heterozygous combination is distinguishable by a phenotype).

distribution function: A graph of some precise quantitative measure of a character against its frequency of occurrence.

distributive DNA replication: *See* DISPERSIVE REPLICATION.

disulphydryl bridge: A particular type of covalent bond, incorporating two sulphur atoms, that can form between two cysteines in a single or on different polypeptides.

ditelocentric: Lines lacking one pair of chromosome arms.

ditelomonotelosomic: An individual in which one chromosome pair is missing but which has a pair of telocentric chromosomes for one arm and an unpaired telocentric chromosome of the other arm of the missing chromosome pair.

ditelosomic: An individual deficient in two homologous arms.

ditelotrisomic: An individual deficient for one chromosome but having a pair of homologous telocentric chromosomes for one arm of the missing chromosome.

ditertiary compensating trisomic: A compensating trisomic in which the missing chromosome is compensated for by two tertiary chromosomes, with one having one arm of the missing chromosome and the other having the second arm. (Kimber and Sears, 1968)

divergence: The percent difference in nucleotide sequence between related DNA sequences or in amino acid sequences between two proteins.

divergent transcription: The initiation of transcription at two promoters facing in opposite directions, so that transcription proceeds in both directions from a central region.

diversifying selection: Selection in which two or more genotypes have optimal adaptiveness in different subenvironments.

divided gene: A gene whose exons are located far away in the chromosome, e.g., a gene for ribosomal protein of smaller subunit (12 rps) in chloroplasts. This gene consists of three exons, with five exons located far away from the other two in all land plants. Mature mRNA from this gene is supposed to be produced by transsplicing.

dizygotic twins: Twins formed from two eggs fertilized at the same time.

D-loop (displacement loop): Configuration found during DNA replication of chloroplast and mitochondrial chromosomes wherein the origin of replication is different on the two strands. The first structure formed is a displacement loop, or D-loop.

DNA: Deoxyribonucleic acid, the genetic material; single- or double-stranded polynucleotide chain.

DNA B helicase: The enzyme that causes unwinding of DNA double helix prior to DNA replication in λ phage and *E. coli*; this enzyme is localized in these systems by proteins λP and DNA C, respectively.

DNA banding: A topological state in which a DNA molecule is bent so that two distant sites on the double helix are brought close together.

DNA binding domain: The part of a transcription factor that tethers the protein to its cognate regulatory element.

DNA bubbles: The nucleic acid configuration during replication in eukaryotic chromosomes, or the shape of heteroduplex DNA at the site of a deletion or insertion.

DNA C: An *E. coli* protein required to localize DNA replication proteins.

DNA circularization: A DNA fragment generated by digestion with a single restriction endonuclease will have complementary 5′ and 3′ extensions (sticky ends). If these ends are annealed and ligated, the DNA fragment will have been converted to a covalently closed circle, thus becoming circularized.

DNA clone: A section of DNA that has been inserted into a vector molecule, such as a plasmid or a phage chromosome, and then replicated to form many copies.

DNA cloning: *See* GENE CLONING.

DNA concatamer: A DNA molecule that comprises a series of smaller DNA molecules linked head to tail. Concatamers are formed during the replication of some viral and phage genomes. (Thomas, 1970)

DNA concatenates: Multiple circular DNA molecules linked in a chain-like arrangement. Concatenates can arise as a result of the normal processes of DNA replication with circular genomes and cannot be separated without strand breakage. Plasmids in bacteria and kinetoplast DNA in trypanosmes frequently form concatenates.

DNA concatenation: A single, large DNA molecule composed of two or more complete phage chromosomes.

DNA crosslinking: Interstrand thymines of DNA form dimers, thus blocking replication.

DNA filter assay: An analytical procedure used to recognize recombinant DNA in cloned cells. The transformed cells only represent a small percentage of the total population of cells present in a culture. The culture is diluted and placed on petri dishes at such a concentration that individual colonies form from each cell. A replicate of the pattern of cells on the plate is taken using a cellulose nitrate filter. The cells are lysed, and the filter is heated to bind the DNA onto the filter in a denatured form in a position equivalent to that of the cell colonies on the petri dish. The filter is then treated with the probe, which combines by DNA hybridization only at the positions of transformed DNA. These are then located using autoradiography or florescent antibodies, as appropriate.

DNA fingerprinting: The technique useful in establishing near-perfect identity of an unidentified body. It is based on the fact that every organism is conferred with DNA variations in form of a variable number of tandem repeats (VNTRs). The DNA profile of an individual is also known as *genetic signatures*. Accuracy of this method is 1-in-75 billion error probability ratio. DNA fingerprinting using microsatellites also called simple sequence repeats (SSRs) has also been utilized both in plants and animals. (Jeffreys, Wilson, and Thein, 1985)

DNA flexibility: DNA flexibility is an important property of DNA, which along with DNA bending is important in packaging, recombination, and transcription. (Goodman and Nash, 1989)

DNA footprinting: A method for determining the sequence specificity of DNA-binding proteins. It utilizes a damaging agent that cleaves DNA at all regions except where the protein is bound to DNA.

DNA G primase: An enzyme that synthesizes the primer required for λ phage and *E. coli* DNA replication; this enzyme is localized in these systems by proteins λP and DNA C, respectively.

DNA glycosylases: Endonucleases that initiate excision repair at the sites of various damaged or improper bases in DNA.

DNA gyrase: An enzyme of *E. coli* that catalyzes negative supercoiling of closed duplex DNA. It is involved in DNA replication, genetic transcription, DNA repair, and genetic recombination. DNA gyrase acts by (a) binding to DNA; (b) introducing supercoils in the presence of ATP; (c) relaxating supercoils in the absence of ATP; (d) causing site-specific cleavage of DNA following denaturing of the enzyme-DNA complex; (e) aiding ATP-dependent catenation and uncatenation of facilitating DNA rings; and (f) facilitating DNA-dependent hydrolysis of ATP to ADP and P_i. (Gellert et al., 1976)

DNA helicase: Any of a class of DNA-unwinding enzymes that either unwind DNA progressively or nonprogressively. Some DNA helicases interact with other replication proteins and are involved in the synthesis of RNA primers on the lagging strand. (Kuhn, Abdel-Monem, and Hoffman-Berling, 1978)

DNA helix A: Right-handed helix; rotation/base pair 33.6°; mean number of base pairs/turn 10.7; inclination of base to helix axis +19°; rise/base pair along helix axis 2.3Å; pitch/turn of helix 24.6Å; mean propeller twist +18°; glycosyl angle confirmation anti; sugar pucker conformation C3′-endo. This form of DNA occurs at high humidity. Also known as *alpha DNA* and *A DNA*.

DNA helix B: Right-handed helix; rotation/base pair 38.0°; mean number of base pairs/turn 10.0; inclination of base to helix axis − 1.2°; rise/base pair along helix axis 0.32Å; pitch/turn of helix 33.2Å; mean propeller twist +16°; glycosyl angle confirmation anti; sugar pucker conformation C1′-endo to C2′-endo. The most commonly found right-handed form of DNA. Also known as *B DNA*.

DNA helix H: DNA containing repeated tracts of pyrimidines and purines such as $(TC)_n$ paired with $(AG)_n$ can form a three-stranded helix in response to negative superhelicity or low pH. The triplex is formed when half of the normal duplex is disturbed and the polypyrimidine strand folds back and inserts itself into the major groove of the remaining duplex. The donated strand associates by *Hoogsteen base pairing* with the purines in the duplex region without disrupting conventional Watson-Crick pairing. The triplex DNA forms a right-handed helix and has a loop of about four nucleotides in the polypyrimidine strand at its tip; the base forms a flexible hinge, hence the synonym *hinge DNA*. Also known as *H DNA* and *triplex DNA*.

DNA helix Z: Left-handed helix; rotation/base pair $-60°/2$; mean number of base pairs/turn 12.0; inclination of base to helix axis $-9°$; rise/base pair along helix axis 3.8Å; pitch/turn of helix 45.6Å; mean propeller twist 0°; glycosyl angle confirmation anti at C, syn at G′; sugar pucker conformation G2′-endo at C, G1′ exo to C1′-exo at G. Also known as *Z DNA*.

DNA insertase: An enzyme that directly inserts missing bases into DNA. The enzyme adds adenine at apurinic sites and guanine at apyrimidinic site. (Lin, Kuhnlein, and Deutsch, 1979)

DNA ligase: An enzyme that repairs single-stranded discontinuities in double-stranded DNA molecules. In the cell, ligases are involved in DNA replication. Purified ligases are used in the construction of recombinant DNA molecules. (Gellert et al., 1976)

DNA looping: A mechanism of gene regulation which imagines that proteins bound with DNA at widely separated sites interact with each other with the intervening DNA looping or bending. It is the interaction between DNA-bound proteins, not the looping per se that regulates gene expression.

DNA methylation: Methylation of adenine of prokaryotic DNA playing a role in host restriction and modification and cytosine in eukaryotic DNA playing a role in the maternal inheritance of chloroplast DNA, DNA replication, recombination, mutation, chromosome folding and packing, mammalian chromosome inactivation, conformational changes in DNA, etc. DNA methylation protects the host DNA from its restriction enzymes. Also known as *DNA modification*.

DNA methylation theory of development: A theory put forward to explain developmental reprogramming by reprogramming of gametes prior to fertilization that depends upon specific *de novo* methylation of chromosomal DNA.

DNA oozing: A gene regulatory mechanism according to which binding of a regulatory protein to its operator helps binding of another protein to adjacent sequences, which in turn helps another protein bind next to it, and so on, until a procession of proteins has oozed out from the control sequence to the point where transcription is initiated.

DNA polymerase: An enzyme that catalyzes the synthesis of DNA on a DNA template from deoxyribonucleotide triphosphate precursors.

DNA polymerase L: The *E. coli* enzyme that completes synthesis of individual Okazaki fragments during DNA replication by degrading the primer and filling in the gaps. May also be helpful in termination of DNA replication.

DNA polymerase II: The main DNA replicating enzyme of *E. coli*.

DNA polymerase III: The enzyme required for *in vivo* DNA replication.

DNA profiling: *See* DNA FINGERPRINTING.

DNA repair: The correction of nucleotide errors introduced during DNA replication or resulting from the action of mutagenic agents.

DNA replication: The process by which a DNA molecule makes its identical copies.

DNA replication fidelity: DNA replication ensures fidelity by keeping mutation rate low. This is achieved through three mechanisms: (a) selection of the correct deoxynucleoside triphosphate (dNTP) substrate from end of the growing chain (base selection); (b) exonucleolytic removal of an incorrectly inserted deoxynucleoside monophosphate (dNMP) from the end of the growing chain (editing); and (c) postreplicative excision of an incorrectly inserted dNMP after the polymerase has extended the DNA chain (mismatch repair).

DNA replication inhibitor: Some compounds bind DNA covalently and inhibit DNA replication. Some well-known DNA replication inhibitors are the following: acridine, acriflavine, 8-aminoquinoline, echinimycin, and proflavine.

DNA sequencing: Determination of the order of nucleotides in a DNA molecule. Methods: (a) *chemical cleavage method* of Maxam and Gilbert (1977); (b) *plus and minus method* of Sanger (1977); and (c) *wandering spot analysis.*

DNA sliding: This mechanism of gene regulation imagines that a protein recognizes a specific site on DNA and then moves (slides, tracks) along the DNA to another specific sequence, whereby interacting with another protein, it initiates transcription. This idea also imagines that the regulatory protein remains bound at the first site and the DNA is thread past or through the bound protein until the second critical site is encountered.

DNA strand: One of the two DNA strands in a double-helical DNA molecule.

DNA synthesis *in vitro*: Synthesis of DNA outside a cell (in a test tube) involving four types of triphosphate nucleosides. A template is used in artificial DNA synthesis.

DNA topoisomerases: The enzymes that introduce or remove turns from the double helix by transient breakage of one or both polynucleotides. DNA topoisomerases are important during transcription. RNA polymerase generates positive supercoiling ahead and leaves negative supercoiling behind. Type I DNA topoisomerase rectifies the situation behind by making a transient break in one strand of DNA. Type II DNA topoisomerase relaxes negative supercoiling during transcription by introducing a transient double-stranded break in DNA. (Wang, 1985)

DNA twisting: A mechanism that explains gene regulation through binding of regulatory proteins to some altered form of DNA (left-handed or single-stranded). An alternate belief is that regulatory protein has an enzymatic activity which alters DNA conformation and regulates gene expression.

DNA unwinding enzyme: An ATP-dependent enzyme involved in the early stages of DNA replication. It unwinds the parental strand to create the two template strands. (Alberts and Frey, 1970)

DNA vector: A system, such as a plasmid or a virus, that is used as a vehicle to transfer DNA from one cell to another.

DNA-binding protein: Any protein that attaches to DNA as a part of its normal function (e.g., histone, RNA polymerase, and *lac* repressor).

DNA-DNA hybridization: When DNA from the same or different sources is heated and then cooled, double helixes will reform at homologous regions. This technique is useful for determining sequence similarities and degree of repetitiveness among DNAs.

DNA-driven hybridization: This process involves the reaction of an excess of DNA with RNA.

DNA-driven reaction: A reaction in which nucleic acid hybridization is utilized in an experimental situation, and complementary DNA is greatly in excess of a radioactive tracer RNA. Therefore, the RNA is forced to compete with a much greater proportion of hybridizing DNA, and thus will bind to highly repetitious DNA sequences.

DNA-mediated gene transfer: The method of gene transfer in which a purified DNA fragment carrying a gene of interest is mixed with a carrier DNA and is precipitated out of solution with calcium phosphate. When target cells are incubated with the precipitate, there is increased uptake of DNA and some of the cells are transformed by the desired gene.

DNA-protein interaction: Interaction between DNA and protein molecules for precise DNA transactions, that is, initiation of DNA replication, initiation of transcription, site-specific recombination, and gene regulation.

DNA-RNA hybridization: When a mixture of DNA and RNA is heated and then cooled, RNA can hybridize (form a double helix) with DNA that has a complementary nucleotide sequence.

DNase: Any enzyme that hydrolyzes DNA.

docking protein: Responsible for attaching (docking) a ribosome to a membrane by interacting with a signal particle attached to a ribosome destined to be membrane-bound.

domain: (1) A discrete structural entity defined as a region within which supercoiling is independent of other domains. (2) An extensive region including an expressed gene that has heightened sensitivity to degradation by the enzyme DNase I.

domestication: A cultivated species, after undergoing many changes due to conscious or unconscious selection such that it no longer exists in a wild form, is said to be domesticated.

dominance: The situation in which one member of a pair of allelic genes expresses itself in whole (complete dominance) or in part (incomplete dominance) over the other member.

dominance hypothesis: The theory that heterosis is caused by the masking of harmful recessive alleles by dominant alleles. *See* OVERDOMINANCE HYPOTHESIS.

dominance modifiers: Those genes that affect the dominance-recessive relationship through relative strength of action of the two alleles at a locus or action of other genes (epistasis).

dominance portion of genetic variance (V_D): The portion of the genetic variance for a given trait resulting from the fact that heterozygotes do not always score exactly midway between the homozygotes.

dominant: Refers to the allele whose phenotypic effect is expressed in a heterozygote. (Mendel, 1866)

dominant epistasis: A condition where a dominant gene masks the expression of nonallelic members.

dominant gene effects: Gene action with deviations from the additive such that the heterozygote is more like one parent than the other.

dominant lethal: Action of a gene lethal in both the homozygous and heterozygous conditions; a gene that frequently kills in a single copy.

dominant phenotype: The phenotype of a genotype containing the dominant allele; the parental phenotype that is expressed in a heterozygote.

dominigenes: Modifying genes that are able to alter the dominance of another gene.

donor plant: The source plant used for propagation, which can be a plant, an explant, a graft, or a cutting. Source plants used for micropropagation are usually pathogen-tested (disease-free) Also known as the *mother plant.*

donor splicing site: *See* LEFT SPLICING JUNCTION.

dosage compensation: A phenomenon in which the activity of a gene is increased or decreased according to the number of copies of that gene in the cell. (Muller, 1932)

dosage effect: A type of inheritance in which two like dominant alleles have a stronger effect than a single one of the same type. (Stern, 1929)

dot blot: A blotting technique of DNA already cloned that eliminates the electrophoretic separation step. Autoradiographs reveal dots rather than bands on a gel, indicative of a cloned gene.

double back cross: The mating of an individual who is heterozygous at two loci with an individual who is homozygous recessive at the same two loci.

double cross: A hybrid between two single crosses of different, pure inbred lines. Most commercial corn hybrids are double crosses.

double crossing-over: Two crossing-over events occurring in a chromosomal region under study. (Sturtevant, 1914)

double digest: The product formed when two different restriction endonucleases act on the same segment of DNA.

double fertilization: In angiosperms, the fusion of one male gamete with the egg nucleus (which forms an embryo) at the same time as a second male gamete fuses with a second female gamete to beget nutritive tissue (endosperm). (Navashin, 1899)

double haploid: A diploid plant produced by doubling the chromosome content of a haploid plant. A doubled haploid will be homozygous at all loci. (Warmke and Blakeslee, 1939)

double helix: The base-paired structure comprising two poly-nucleotides that is the natural form of DNA in the cell.

double isocompensating trisomic: An individual deficient in one chromosome, which is compensated by two isochromosomes, one for each arm of the missing chromosome.

double monoisosomics: An individual missing one chromosome pair but having two isochromosomes, one for each arm of the missing pair. (Kimber and Sears, 1968)

double monotelosomic: An individual in which one chromosome pair is missing but two telocentrics, one for each arm of the missing pair, are present. (Kimber and Sears, 1968)

double recombination: Two recombination events between the same DNA molecules, which may result in insertion of a short linear molecule into a longer linear or circular molecule.

double reduction: The condition in polyploids where segments of sister chromatids remain together at the end of the first as well as at the end of the second meiotic division and enter the same gamete. (Darlington, 1929b)

double telotrisomic: An individual deficient in one chromosome but having two telocentric chromosomes, one for each arm of the missing chromosome. This is really a *pseudotrisomic*.

double tertiary compensating trisomic: A compensating trisomic in which a missing chromosome is compensated by two trans-located chromosomes.

doubling dose: The dose of a mutagenic substance to induce as many mutations as occur spontaneously in one generation.

down promoter mutations: A promotor that decreases the frequency of initiation of transcription.

downstream: Toward the 3′ end of a polynucleotide.

drastic amino acid substitutions: When amino acids having dissimilar charges, chemically reactive and nonreactive, or bulky and light-side chain amino acids replace each other.

driver DNA: Excess DNA used in a DNA-driven reactions.

drug resistance gene: A gene that codes for an enzyme that enables the host cell to resist the effects of a drug. In general, the enzyme hydrolyzes the drug or modifies its structures in some way. Drug resistance genes include the genes conferring antibiotic resistance that are carried on the plasmids.

ds DNA: Double-stranded DNA.

duplex: A double-stranded nucleic acid molecule or a double-stranded region of a mainly single-stranded molecule. (Blakeslee, Belling, and Farnham, 1923)

duplicate dominant epistasis: Dominant alleles at either of two loci can mask the expression of recessive alleles at the two loci resulting in a 15:1 ratio. Also called *duplicate epistasis*.

duplicate genes: Either dominant or both dominant genes together produce the same phenotype to give a 15:1 ratio in F_2 progeny. (Bridges, 1919)

duplicate loci: A pair of nonallelic loci having the same genetic meaning whether alone or together; two isoloci. (Shull, 1914)

duplicate recessive epistasis: Recessive alleles at either of the two loci can mask the expression of dominant alleles at the two loci resulting in a 9:7 ratio. Also called *complementary epistasis*.

duplication: A chromosomal aberration in which a segment of the chromosome bearing specific loci is repeated.

durable resistance: The resistance that remains effective in a cultivar during its widespread cultivation for a long sequence of generations or a period of time, in an environment favorable to disease or pests. Such resistance is controlled by several genes. Both recognizing durable resistance and understanding genetic components related to durability are essential requirements in breeding for durable resistance.

dyad: Two sister chromatids attached to the same centromere.

dysgenic: Any effect or situation that is or tends to be harmful to the genetics of future generations.

dysploidy: Different diploid number of species in a genus, other than a polyploid species. (Tischler, 1950)

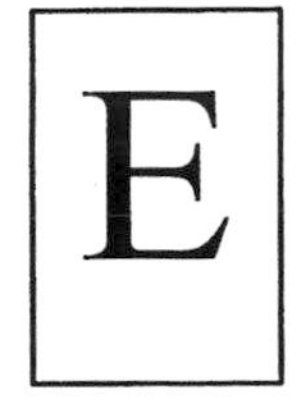

early gene: Genes contained in the circular genomes of viruses such as bacteriophage T_4 and SV40. Early genes are transcribed early in the viral replication cycle, and some of their products are necessary for the transcription or translation of other genes in the genome, which are termed late genes in double-stranded DNA viruses. The early genes tend to be clustered on one DNA strand and late genes on the other; transcription involves both strands, each of which is read in an opposite direction.

early generation test: Discarding poor recombinants early in the segregating generations (F_2 or F_3) so as to pay adequate attention to potential recombinants later.

Eco RI restriction enzyme: A type II restriction enzyme from *E. coli* RY13. Eco RI recognizes the DNA sequence shown below and cuts at the sites indicated by the arrows.

```
      ↓
5′  G A A T T C  3′
3′  C T T A A G  5′
              ↑
```

ecological genetics: A branch of genetics that analyzes evolution in a natural population and examines the adjustments and adaptations of wild populations to their environment.

economic heterosis: Superiority of the F_1 hybrid over the standard commercial cultivar. Also referred to as *heterobeltiosis* or *standard heterosis*.

ecophene: The range of phenotypes produced by one genotype within the limits of habitat under which it is found in nature. (Turesson, 1922)

ecophenotype: A nongenetic modification of the phenotype in response to an environmental condition.

ecospecies: A group of species defined as being so related as to be capable of exchanging genes freely (that is, of hybridizing) without loss of vigor or fertility in the progeny. (Turesson, 1922)

ecotype: A local race that owes its most conspicuous attributes to selective effects of a local environment.

ectopic gene expression: The expression of a gene in a tissue in which it is normally inactive. This is normally achieved by linking the coding region of the gene to a heterologous promotor.

edible vaccines: A gene that produces a product of medicinal value is transferred into a plant that expresses the product only in the edible part.

edition enzyme: Enzyme responsible for proofreading the DNA and correcting errors in replication or following from mutation.

effective breeding size: The breeding size of a population, adjusted mathematically to permit comparisons with others. (Wright, 1931)

effective population size (N^C): The number of individuals in a population corrected for biological factors, such as the sex ratio, that allows different populations to be directly compared with respect to genetic drift.

effector: A molecule that influences the behavior of a regulatory molecule, such as a repressor protein, thereby influencing gene expression. (Jacob and Monod, 1961a)

effector gene: The gene that drives transcription of another gene with the help of a promoter from another gene.

effector molecule: A molecule (a sugar, an amino acid, or a nucleotide) that can bind to a regulator protein and thereby change the ability of the regulator molecule to interact with the operator.

efficiency elements: Eukaryotic gene regulatory elements that act to increase the overall rate of genetic transcription but do not directly confer tissue specificity of expression.

egg cell: The female gamete, especially in plants.

electro-bolting transfer: The electrophoretic transfer of macromolecules (DNA, RNA, or protein) from gel in which they have been separated to a support matrix such as a nitrocellulose sheet. This is an alternative to the capillary transfer usually used in techniques such as Southern and Northern blotting.

electromorphs: Allozymes that can be distinguished by electrophoresis. (King and Ohta, 1975)

electron microscope: A powerful microscope that uses beams of high-speed electrons instead of light waves.

electrophoresis: Separation of molecules on the basis of their mass and net electrical charge ratio.

electrophoretic karyotype: Due to the very small size of yeast chromosomes (260 to 2,000 kb) and absence of their compact organization, these cannot be resolved under the compound light microscope. Therefore, normal karyotype of these chromosomes cannot be constituted. Different chromosomes of yeast can be separated by pulse-field gel electrophoresis, in which chromosome size DNA molecules are visualized as ethidium-bromide-stained bands. This constitutes an electrophoretic karyotype. (Carle and Olson, 1985)

electroporation: Creation, by means of an electrical current, of transient pores in the plasmalemma, usually for the purpose of introducing exogenous material, especially DNA, from the medium. A technique for transfecting cells by the application of a high-voltage electric pulse.

element: The particulate hereditary determiner of Mendel. *See* GENE.

elicitor-receptor model: A model proposed to explain gene-for-gene hypothesis. Accordingly, either primary virulence gene products or metabolites resulting from their metabolic activity are recognized by specific plant receptors encoded by disease genes.

elicitor-suppressor model: A model proposed to explain gene-for-gene hypothesis. Accordingly, pathogens are thought to produce general elicitors that initiate host defense unless a specific suppressor substance is also liberated by a particular pathogen. This scheme is understood to work where virulence rather than avirulence is dominant.

ELISA (enzyme-linked immunosorbent assay): A sensitive immunological assay system that avoids both the hazards and expense of radioactive or fluorescence detection system. Two antibody preparations are used in ELISA. The primary antibody binds the antigen and is itself bound by the second, antibody, antiglobin. The antiglobin is linked to an enzyme, e.g., horseradish peroxidase, whose activity is easily monitored, for instance by a color change. The extent of the enzymatic reaction is then a quantitative indication of the amount of primary antibody or, indirectly, of antigen present.

elution: Washing out or collection of liquid from a chromatography column. The liquid, which may contain the fractionated samples, is known as the *eluate*.

emasculation: Removing the anthers from a bud or flower before pollen is shed. Emasculation is a normal preliminary step in crossing to prevent self-pollination.

embiont: A heterotrophic protocell that through the evolution of metabolic pathways became autotrophic.

embryo: A young organism in the first stages of development; the immediate product of fertilization.

embryo culture: *In vitro* development or maintenance of isolated mature or immature embryos.

embryo sac: A large, thin-walled cell within the nucleus of the ovule of seed plants, in which the egg and the embryo develop.

embryogenesis: The process of embryo initiation and development.

embryoid: Nonzygotic embryo formed in culture.

emigration: Outgoing of alleles from a population.

end-labeling: The addition of a radioactively labeled group to one end ($5'$ to $3'$) of a DNA strand.

endomitosis: An increase in somatic DNA content, which takes place within an intact nuclear envelope and gives rise endopolyploidy. (Geitler, 1939)

endonuclease: An enzyme that breaks phosphodiester bonds within a nucleic acid molecule.

endoplasmic reticulum: A double-membrane system in the cytoplasm, continuous with the nuclear membrane and bearing numerous ribosomes.

endopolyploidy: An increase in the number of chromosome sets caused by replication without cell division.

endoreplication: Replication without separation of chromatids.

endosperm: A polyploid (in many species, triploid) food storage tissue in many angiosperm seeds formed by fusion of two (or more) female cells and a sperm.

endosperm mutation: A change that is detected by a difference in the endosperm of the corn kernel, e.g., starchy to sweet.

endosymbiont: The organism that resides inside a host organism. The term also applies to those organisms that reside within the cells of an other organism.

enhancer: A special type of eukaryotic regulatory sequence that can increase the rate of transcription of a gene located some distance away in either direction. They are mostly *cis*-acting, but transacting ones are also known. They are effective whether lying upstream or downstream. They are active whether they lie in same or opposite polarity as the gene. They are equally effective regardless of the organism or the gene from which they are derived when attached to foreign DNA.

enucleate cell: A cell having no nucleus.

environment, external: The aggregate of all external conditions and influences affecting the life and development of an organism.

environment, internal: The cellular environment affecting the expression or manifestation of certain genes, e.g., hormonal constitution affects expression of the gene for baldness in men and women.

environmental mutagenesis: *See* GENETIC TOXICOLOGY.

environmental variance: The variance due to environmental variation.

environmental variation: When individuals with the same genotypes develop in different environments, their phenotypes may be quite different. For example, identical twins raised in different environments show different phenotypes due to environmental variation.

enzyme: Any substance, protein in whole or in part, that regulates the rate of a specific biochemical reaction in living organisms.

enzyme kinetics: The rate of a reaction controlled by an enzyme. In low substrate concentrations the reaction rate (V) is always proportional to the substrate concentration ([S]), but when the substrate is present at high concentrations, enzyme saturation will occur (Vmax). The Michaelis-Menten equation expresses the rate of a reaction as V = Vmax [S]/(Km+[S]), where Km is the Michaelis constant and is equal to the concentration of the substrate when the speed of the reaction is half its maximum.

enzyme regulation: Enzymes may be regulated at two levels: (a) at the level of gene expression and protein synthesis through induction and repression; (b) at the enzyme level through enzyme inhibition or activation as a result of the binding of effector molecules at allosteric sites on the enzyme.

enzyme repression: A mechanism that prevents the synthesis of an enzyme by the formation of repressors that bind to DNA, thus preventing transcription.

ephemeral resistance/performance: Impressive resistance or performance when the variety is new or in breeder's trials, but it fades when the variety is released or becomes popular (under commercial cultivation).

ephestia: *See* MATERNAL INFLUENCE.

epigenesis: A concept according to which development of an individual is epigenetic and genes act in epigenetic systems and adult structures develop from uniform embryonic tissues.

epigenetic: Nongenetic causes of a phenotype. (Waddington, 1942)

epigenetic event: Any change in a phenotype that does not result from an alteration in DNA sequence. This change may be stable and heritable and includes alteration in DNA methylation, transcriptional activation, transitional control, and posttranslational modifications.

epigenetic modifications: Changes that influence the phenotype without altering the genotype. *See* POSTTRANSLATIONAL MODIFICATIONS.

epigenetic variation: Phenotypic variability that has a nongenetic basis.

epigenotype: The total developmental system; the totality of interactions among genes resulting in the phenotype. (Waddington, 1939)

epimutations: Heritable changes in gene activity due to DNA modification.

epiphytotic: Sudden and usually widespread development of a destructive disease in plants.

episome: A plasmid capable of integration into the host cell's chromosome. These are those particles that are added to a genome through an external source—not by mutation or rearrangement of the existing genome. (Thompson, 1931; Jacob and Wollmann, 1958)

epistasis: The ability of one gene to mask the phenotype derived from a second gene. (Bateson, Saunders, and Punnett, 1908)

epistasy: The property of nonreciprocal conditioning of the manifestation of one gene difference, said to be hypostatic, by the action of another, said to be epistatic. *See* EPISTATIC GENE and HYPOSTATIC GENE.

epistatic: A gene whose effect overrides that of another gene with which it is not allelic.

epistatic genes: The genes that suppress expression of other genes.

epistatic variance: What remains unaccounted for by additive and dominance variance components in genetic variance, i.e., failure of the summation of the intralocus genetic variance to account for the total variation among genotypes.

equational (homotypic) division: The second meiotic division is an equational division because it does not reduce chromosome numbers. (Weismann, 1887)

equatorial plate: The figure formed at the spindle equator in nuclear division. (Van Benden, 1883)

equilibrium gene frequency: In a population, the gene frequency that varies nondirectionally about a mean by an amount described by the standard deviation under conditions of unchanging selection pressure and mutation rate, with no intermixing from other populations.

error variance: Variance arising from unrecognized or uncontrolled factors in an experiment with which the variance of recognized factors is compared in tests of significance.

error-catastrophy model of aging: Mutations in the protein-synthesizing apparatus have a cascading effect on a widening series of cell products, causing cell malfunction. If an error occurs in the proteins that are needed for the synthesis of other proteins, a wrong enzyme will synthesize many more wrong proteins before it has degraded itself. Consequently, the activity of the cell gets so slow that it dies due to "error catastrophe."

error-prone DNA repair: A type of DNA repair in which DNA chain growth occurs across damaged DNA segments, but some of these new strands are themselves defective. Such error-prone repair appears to be the chief source of genetic mutation in the bacteria in which it occurs. *See* SOS RESPONSE.

essential gene: A gene whose deletion is lethal to the organism.

established cell line: A line which consists of eukaryotic cells that have been adapted to indefinite growth in a culture (they are said to be immortalized).

estimation of linkage: Calculating distance between genes form F_2 or TC_1 progeny on the basis of crossing-over frequency.

ethidium bromide: A molecule that can intercalate into DNA double helixes when the helix is under torsional stress.

ethionine: A methionine analogue that lacks its methyl group and thereby inhibits most methyl transferases, thus inducing gene expression in eukaryotes by undermethylation of cytosine.

euchromatin: The regions of a eukaryotic chromosome that are less condensed and stain less deeply with DNA-specific dyes. These are composed of 100 Å chromatin. (Heitz, 1929)

eukaryote: An organism whose cells are characterized by the presence of membrane-bound nuclei. (Chatton, 1925)

euploid: An individual having changes by complete sets or exact multiples of the monoploid, e.g., triploid, tetraploid, and so on. Euploids above the diploid level may be referred to collectively as *polyploids*. (Täckholm, 1922)

evolution: The transformation of an organism in a way that descendants differ from their predecessors.

evolution by random walk: A hypothesis that belives in evolution of species through forces acting at random rather than in a specific direction.

evolutionary breeding: Breeding procedure in which the variety is developed from an unselected progeny of a cross, or multiple crosses, that has undergone evolutionary changes.

evolutionary clock: This is defined by the rate at which mutations accumulate in a given gene.

evolutionary rates: The rate of divergence between taxa, measurable as amino acid substitutions per million years.

evolutionary responses: Interacting populations form an environment for each other and mutually impose selective forces that stimulate evolutionary changes to alter their relationship by making all the interacting populations respond.

evolutionary species concept: In this species concept, emphasis is placed on different evolutionary lineages. Also known as *phylogenetic species concept.*

ex vitro: Organisms removed from a culture and transplanted generally to soil or a potting mixture.

exaggeration: Expression of a hypomorphic gene placed opposite to a deficiency. (Mohr, 1923)

excision: The enzymatic removal of a nucleotide or polynucleotide fragment from a nucleic acid polymer.

excision repair: DNA repair that results in the removal of a defective base and adjacent sequence by excision and filling the gap by repair, replication, and ligation.

exclusion principle: The principle stating that two species cannot coexist at the same locality if they have identical ecological requirements.

exon: One of the coding regions of a discontinuous gene. More precisely, the sequence present in a gene which is complementary to that present in its mRNA. (Gilbert, 1978)

exon shuffling: A theory that supposes exons to contain discrete subunits of genetic information that may be "shuffled" to produce new functional genes.

exon theory of genes: Split genes arise not by insertion of introns into unsplit genes but from combinations of primordial "mini-genes" separated by spacers.

exonuclease: An enzyme that sequentially removes nucleotides from the ends of a nucleic acid molecule.

expected ratio: In statistical evaluation of an observed ratio, the ratio by which deviations are calculated to determine whether they are significant.

experimental alteration of germ plasm: The changing of the germ plasm so that the inheritance is altered. It can be produced by ionizing radiation, ultraviolet light, heat, chemicals, and such novel methods as transduction and transformation.

experimental design: A branch of statistics that attempts to outline the way in which experiments should be carried out so the data gathered will have statistical value.

explanation: Preparation of an explant by isolating tissue or a group of cells from an organism or embryo.

explant: The plant part that is used for regeneration in tissue culture.

explant culture: The maintenance or growth of an explant in a culture.

exploitation competition: A form of competition that involves the superior ability to gather resources, rather than an active interaction among organisms for these resources.

exploitive adaptation: Genetic capacity of an organism to bring out phenotypic changes in response to an altered environment so as to achieve acclimatization.

explosive speciation: Evolution of a large number of species from a single closed species.

exponential growth: A phase of growth of a culture or a population in which the time of doubling in size (or more generally, the population growth rate) is constant. Under exponential growth, the logarithm of the number of individuals increases linearly with time. Also known as *geometric growth* or *logarithmic growth.*

expressed sequence tags (EST): It is polymerase chain reaction (PCR)–based restriction fragment length polymorphism (RFLP) that uses the end sequences of a cDNA clone as a primer.

expression screening: A method of screening for a specific cDNA clone where the cDNA is inserted next to a promotor active in the host cell and an immunoassay or bioassay is used to detect the required clone.

expression vector: A vector that results in the expression of inserted DNA sequences when propagated in a suitable host cell, i.e., the protein coded by the DNA is synthesized by the host's system.

expressivity: The degree of phenotypic expression within one phenotype under a variety of environmental conditions. Some genes show *variable expressivity.* (Vogt, 1926)

expressivity modifiers: Those genes that are responsible for variable expressivity of a phenotype.

extended anticodon hypothesis: The structure of an anticodon loop and the proximal anticodon stem are related to the sequence of the anticodon. Thus, anticodon is extended to (a) two nucleosides at the 5′ end of the anticodon, (b) three nucleosides of the anticodon, and (c) five pairs of nucleosides in the anticodon stem. Extended anticodons are involved in translation efficiency.

extrachromosomal gene: Any gene that is not carried by the cell's chromosome(s), for example, genes present on mitochondrial or chloroplast genomes or genes carried by plasmids.

extranuclear genes: The genes that reside in organelles such as mitochondria and chloroplasts outside the nucleus.

extranuclear inheritance (cytoplasmic inheritance): This is characterized by differences in reciprocal crosses, progeny shows characters of the female parent, and there is no linkage with the nuclear genes.

eyes: Referring to the configuration of replicating DNA in eukaryotic chromosomes.

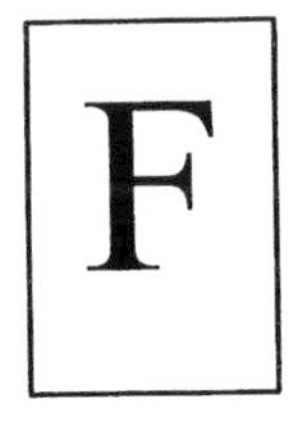

F_1, F_2, etc: Refer to first and second filial generations, and so on.

F plasmid: A conjugative circular plasmid mediating the *E. coli* mating process. It denotes fertility since it can transfer not only itself from one bacterial strain to another, but it can mobilize portions of the bacterial chromosome. Also known as *F factor* or *F sex factor.*

factor: Mendelian unit of inheritance. *See* ELEMENT and GENE. (Mendel, 1866)

factorial experiment: One in which all the treatments of agents under investigations are varied simultaneously and combined in such a way that any derived effect of one or a group of them may be isolated and separately evaluated.

facultative heterochromatin: The inert state of sequences that also exist in active copies.

facultative pathogen: A microorganism that is capable of growth and reproduction in either a nonliving medium or a living host; in the latter instance, a disease condition of the host may arise. Compare OBLIGATE PATHOGEN.

family selection: A breeding technique of selecting a pair on the basis of the average performance of their progeny.

farmer's fixative: (1) A ratio of 3 anhydrous ethanol to 1 glacial acetic acid; a fixing and dehydrating agent used in histology. (2) Used in conjunction with Feulgen stain and carbolfuchsin stain for chromosome analysis.

fast component of DNA: The component of a reassociation reaction that is the first to renature and contains highly repetitive DNA.

fatuoid: A mutant commonly occurring in cultivated oats and that resembles wild oats *(Avena fatua)*.

favored mutations: These are missense mutations that affect less critical sites of a polypeptide chain such that the basic functional property of the polypeptide does not change. Such mutations may, however, affect some kinetic properties of the protein/enzyme and may be more efficient under one environment or the other. Such mutations are favored by the process of evolution.

feedback: The influence of the result of a process upon the functioning of the process. (Warburton, 1955)

feedback inhibition: A posttranslational control mechanism in which the end product of a biochemical pathway inhibits the activity of the first enzyme of this pathway. (Yates and Pardee, 1956)

feeder layer: A layer of cells formed by density-dependent stimulation of cells growing in a culture.

fertility: Reproductive potential measured by the total quantity or relative proportion of developing eggs or of matings resulting in progeny.

fertility factor: A bacterial fertility factor, F, that carries a portion of the bacterial chromosome. Also known as *F factor. See* F PLASMID. (Lederberg, Cavelli, and Lederberg, 1952)

fertility restoration: Restorer gene restores fertility in cytoplasmic male sterile lines.

fertility-restoring genes: Cytoplasmic male sterile lines will produce viable pollen in certain genotypes. Restorers have nuclear genes that restore fertility to a cytoplasmic male sterile line.

fertilization: The fusion of male and female gametes.

Feulgen stain: An important cytochemical test for DNA that utilizes Schiff's reagent and gives a purplish-pink color when positive.

Feulgen test: It is a chemical test for DNA. The test is made by first hydrolyzing the chromosomes with the Schiff's reagent and then treating it with Feulgen stain. It forms a characteristic coloration (magenta or deep purple).

fidelity of DNA replication: Deals with various processes responsible for faithful replication of DNA through generations on dividing cells. The processes recognized in this process are base selection, proofreading, and mispair correction. (Echols and Goodman, 1991)

field resistance: The term used to describe a complex type of resistance that gives partial but effective control of a parasite under natural conditions in the field; however, this resistance is difficult to detect in laboratory or greenhouse tests. Field resistance is often polygenically controlled and durable.

figure eight: Two circles of DNA linked together by a recombination event that has not yet been completed.

filial generation: Any generation of offspring following the parental generation.

filter hybridization: Hybridization performed by incubating a denatured DNA preparation immobilized on a nitrocellulose filter with a solution of radioactivity labeled RNA or DNA.

fine structure genetic mapping: *See* FINE STRUCTURE OF GENE.

fine structure of gene: Deals with analysis of structure of gene using its property of undergoing mutation and recombination. This analysis makes use of *cis-trans* complementation test. *See* *CIS-TRANS* COMPLEMENTATION TEST.

fingerprinting: A technique in which protein or DNA is subjected to enzyme digestion and the resulting fragments are separated by a combination of paper chromatography and electrophoresis, thus revealing a pattern. *See* DNA FINGERPRINTING and PROTEIN FINGERPRINTING.

first division segregation (FDS): In haploid eukaryotes, the segregation of two alleles at a locus at meiosis I. FDS gives 1:1 ratio during tetrad analysis.

fitness: The number of offspring left by an individual, often compared with the average of the population or with some other standard, such as the number left by a particular genotype.

5′ (five prime) and 3′ (three prime) noncoding regions: Untranslated regions at the 5′ and 3′ ends of mRNA sequences.

5′ DNA end: In the direction of the exposed 5-carbon of the deoxyribose of the DNA chain. (At one end of the chain the 5-carbon is exposed; at the other the 3-carbon is exposed.)

5′ P DNA strand terminus: The end of a polynucleotide that terminates with a mono-, di-, or triphosphate attached to the 5′ carbon of the sugar.

5′ to 3′ DNA growth: The synthesis of a nucleic acid chain by joining the 5′ (PO_3) end of a nucleotide to the 3′ (OH) end of the last nucleotide already in the uncompleted chain.

5-azacytidine: A cytosine analogue with a nitrogen atom replacing the carbon atom at position at position 5 of a pyrimidine ring, thus unable to accept a methyl group. A short treatment with it produces foci of differentiated cells lined by undermethylation of cytosine, presumably by inhibiting methyl transferase.

5-bromodeoxyuridine: Growing cells readily incorporate this base analogue into nucleic acids and can then be killed by exposure to light. It suppresses the utilization of deoxythymidine. This procedure is a selection mechanism utilized in mutagenic studies to eliminate wild-type cells and recover the auxotrophs.

5-bromouracil: A base analogue used as a mutagen. This compound can be incorporated into DNA sequences in place of thymine and can cause incorrect base pairing some time latter.

fixable variance: The genetic variance component arising from additive gene action including additive × additive and its higher interactions. This variance helps selection to be most effective.

fixation of alleles: The establishment of a single allele within a population through random chance loss of the other member or members of the same set.

fixative: A compound that stabilizes or fixes other compounds or structures securely so that their structural integrity is retained; the process of fixation utilizes chemical agents to permanently prepare cells or tissues for microscopy. *See* FARMER'S FIXATIVE.

fixed allele: An allele for which all members of the population under study are homozygous, so that no other alleles for this locus exist in the population.

fixed breakage point: According to the heteroduplex DNA recombination model, the point from which unwinding of the DNA double helixes begins, as a prelude to formation of heteroduplex DNA.

fixed rearrangement: When a chromosomal rearrangement is present in different species, it is called fixed.

fixity of species: The axiom that all species, once determined, remain constant (opposite to the theory of organic evolution).

fixity-of-species doctrine: An old doctrine which argues that species were specially created and do not undergo any changes over time.

flanking sequence: A sequence that is either upstream or downstream, on the same DNA molecule, from a particular definitive sequence such as a gene. Enhancer sequences frequently flank gene-coding sequences.

flexible populations: These populations react by genetic changes to seasonal as well as altitudinal environmental alterations.

floating chromosomal rearrangements: When a chromosomal rearrangement is present within a species, it is called floating or polymorphic.

flow cytometry: The method used to sort out whole cells or metaphase chromosomes. Fluorochrome-stained chromosomes are sent through a narrow laser beam and then through a very sensitive flow sorter, which directs chromosomes having different fluoro signals into different chambers.

flow-sorted chromosomes: Chromosomes sorted on the basis of size, by flow cytometry.

fluctuation test of mutations: An experiment by Luria and Delbrück (1943) that compared the variance in number of mutations among small cultures with subsamples of a large culture to determine the mechanism of inherited change in bacteria.

fluorescence *in situ* hybridization (FISH): In this *in situ* hybridization technique, nonradioactive immunofluorescent probes are used for detection of hybridization. The main advantage over conventional *in situ* hybridization is that it is safe and more than one sequence can be visualized in a single hybridization experiment.

fluorescence microscopy: Microscopy in which the microorganisms are stained with fluorescent dye and observed by illumination with ultraviolet light.

fMet: N-formylmethionine, the modified amino acid carried by the tRNA that initiates translation in bacteria.

foldback DNA: Consists of inverted repeats that have renatured by intrastrand reassociation of denatured DNA. (Britten and Kohne, 1970)

footprinting: A technique to determine the length of nucleic acid in contact with a protein. While in contact, the free DNA is digested; the remaining DNA is then isolated and characterized.

forbidden mutations: The mutations that affect the function of a gene so drastically that the individual cannot tolerate them, e.g., frame shift mutations. Such mutations do not accompany the process of evolution.

forces of evolution: Mutation, recombination, migration, random genetic drift, the Founder effect, bottlenecks, assortative mating, selection, etc., are regarded as forces of evolution.

foreign DNA: DNA that is not found in the normal chromosome complement of the organism concerned. It may be introduced into a vector by complexing with native DNA using techniques of gene manipulation, and hence entering the new host. Foreign DNA may also be introduced into host cells following a viral infection.

forward mutation: Any change away from the standard (wild-type).

foundation seed: Seeds of a released variety produced from a breeder's seed or under the direct supervision of a breeder at the experimental station or recognized seed farms.

four-strand stage crossing-over: Crossing-over between chromatids at the diplotene stage in meiotic prophase when four chromatids are present.

fragile chromosome site (FS): Any of the break-prone sites in chromosomes that fail to compact during mitosis and appear as nonstaining gaps.

frameshift intergenic suppressors: A mutation in an RNA gene suppresses mutagenic effect of a frameshift mutation.

frameshift mutation: A mutation that results from insertion or deletion of a group of nucleotides that is not a multiple of three, therefore changing the frame in which the altered gene is translated. Also known as *phaseshift mutation*. (Brenner et al., 1961; Crick et al., 1961)

frameshift suppression: The reversion of the effects of a frame-shift mutation by a second mutation in the same gene, thus restoring the reading frame.

freeze preservation: Conditioning and preservation of plant cells, tissues, or organs at extremely low temperatures, usually in liquid nitrogen.

freeze squeeze: A method of recovering DNA from an agarose gel. The portion of the gel containing the desired class of molecules is cut out and frozen. It is then placed between two sheets of Parafilm TM and squeezed until a drop of DNA solution is ejected. It is claimed that up to 60 percent of the DNA contained in the gel fragment may be recovered in this way.

frequency-dependent fitness: Fitness differences whose intensity changes with changes in the relative frequency of genotypes in the population.

frequency-dependent selection: A selection whereby a genotype is at an advantage when rare and at a disadvantage when common.

frequency-independent fitness: Fitness that is not determined by interactions with other individuals of the same species.

frequency-independent selection: Selection in which the fitnesses of genotypes are independent of their relative frequency in the population.

friability: A term indicating the tendency for plant cells to separate from one another.

frozen-accident theory of genetic code: A theory given by Crick to explain the relationship between an amino acid and its codon. There was a chance selection of code words and then freezing of the selection of code words. Once this association took place, it perpetuated as such. This theory can explain the particulate nature of the codon but not the pattern of degeneracy. This theory was modified by Crick himself. Accordingly, choice of the first codon was made by accident but choice of synonymous codons is made by selection. This explains the degeneracy of code.

functional alleles: Alleles determined on the basis of a complementation test.

functional proteins (enzymes): These serve to catalyze numerous chemical reactions to keep the organism alive.

fundamental number: The number of chromosomes arms in a somatic cell of a particular species. (Matthey, 1945)

fusion gene: A gene derived from parts of two other genes, either by unequal crossing-over, by deletion of part of the chromosome segment between two adjacent genes, or by recombinant DNA technology. The proteins produced by fusion genes are called fusion proteins.

fusion of species: Complete breakdown of reproductive isolating barriers between species. Consequently, some well-defined species merge into one species.

fusion protein: A polypeptide made from a recombinant gene that contains portions of two or more different genes. The different genes are joined so that their coding sequences are in the same reading frame.

G proteins: These are guanine nucleotide-binding trimeric proteins that reside in the plasma membrane. When bound by GDP, the trimer remains intact and is inert. When the GDP bound to the α subunit is replaced by GTP, the α subunit is released from the ßγ dimer. One of the separated units (either the α monomer or the ßγ dimer) then activates or represses a target protein.

Gl phase: The first gap period between the division phase and the chromosome replication phase (S phase) of the cell cycle.

G2 phase: The second gap period between the S phase and the division phase of the cell cycle.

GAL4: A sequence-specific DNA-binding protein that activates transcription in yeast.

gall: A tumorous growth in plants.

gamete: A reproductive cell, usually carrying the haploid chromosome complement, that can fuse with a second gamete to produce a new cell during sexual reproduction. (Strasburger, 1877)

gametic (tissue or generation): Having a number of chromosomes (haploid) in contrast to zygotic tissue with 2n (diploid). Some microorganisms for the most part are haploid.

gametic mutation: A mutation occurring during gamete formation.

gametic selection: The forces acting to cause differential reproductive success of one allele over another in a heterozygote.

gametoclonal variation: Variation among the plants obtained from anther/pollen culture.

gametoclone: Plants regenerated from cell cultures derived from meiospores, gametes, or gametophytes.

gametogenesis: Formation of the gametes; microsporogenesis and megasporogenesis in plants; oogenesis and spermatogenesis in animals.

gametophyte: The haploid stage of a plant life cycle that produces gametes (by mitosis). It alternates with a diploid, sporophyte generation.

gametophytic differences: Differences observable in the gametophyte, e.g., waxy versus starchy pollen in corn and in some other cereals. Most, if not all, of the differences in lower forms (haploid) are gametophytic.

gametophytic incompatibility: A botanical phenomenon controlled by the complex S locus in which a pollen grain cannot fertilize an ovule produced by a plant that carries the same S allele as the pollen grain. For example, S_1 pollen cannot fertilize an ovule made by an S_1/S_2 plant.

GAP: A missing section on one of the strands of a DNA duplex. The DNA will, therefore, have a single-stranded region.

GAP period: One of two intermediate periods within the cell cycle. GAP 1 occurs between mitosis and the DNA synthesis phase; GAP 2 occurs between DNA synthesis and next mitosis.

gas-liquid chromatography (GLC): A chromatographic method in which the materials to be analyzed are vaporized and carried in an inert gas (e.g., nitrogen) through a stationary liquid phase. The method is suitable for analyzing complex fatty-acid mixtures that are volatile at reasonably low temperatures.

Gause's Law: An ecological principle stating that different species occupying identical ecological niches (i.e., ecological homologs) cannot coexist in the same habitat. (Gause, 1934)

G-banding: The chromosome preparations on a slide are air-dried, subjected to denaturation-renaturation and then stained with Giemsa. The characteristic bands produced are called G bands. These bands are generally produced in A/T-rich, heterochromatic regions.

GC box: A component of the nucleotide sequence that makes up the eukaryotic promoter.

gel electrophoresis: Electrophoresis performed in a gel matrix so that molecules of similar electrical charge can be separated on the basis of size.

gel retardation analysis (GRA): A technique that identifies a DNA fragment that has a bound protein by virtue of its decreased mobility during gel electrophoresis.

gene: A segment of DNA that contains biological information and codes for an RNA and/or polypeptide molecule. Recognized through its variant forms that transmit specification(s) from one generation to the next. (Johannsen, 1909)

gene action: The manner in which genes control phenotypic expression of various characters.

gene amplification: A process or processes by which the cell increases the number of a particular gene within the genome. (Brown and David, 1968)

gene arrangement: A particular distribution of genes along a chromosome, typically identified by the distribution of bands along a polytene chromosome.

gene bank: Collection of world germ plasm containing a large fraction of the total variability of a crop including its immediate relatives. Also referred to as *treasure house of variability* maintained either *in situ (gene sanctuary)* or *ex situ (in vivo/in vitro). See* GENE LIBRARY.

gene cloning: Insertion of a fragment of DNA, containing a gene, into a cloning vector and subsequent propagation of the recombinant DNA molecule in a host organism.

gene cluster: A group of adjacent genes that are identical or related.

gene complement: The set of genes characteristic of any one species of organism.

gene concept: Understanding a gene in terms of its structure and function.

gene construct: A special package of genes containing all the information necessary for their expression.

gene conversion: In Ascomycete fungi a 2:2 ratio of alleles is expected after meiosis, yet a 3:1 ratio is sometimes observed. The mechanism of gene conversion is explained by repair of heteroduplex DNA produced by the Holliday model of recombination. (Winkler, 1930; Lindergren, 1953)

gene disruption: A method of inactivating a gene within a living cell by transforming it with a construct which is able to undergo homologous recombination.

gene dosage: The number of times an allele is present in a particular genotype.

gene eviction: A method, originally developed in yeast molecular biology, that permits the retrieval of a chromosomal copy of any gene which has previously been cloned. Thus, if the wild-type copy of a gene has been cloned, gene eviction may be used to obtain a mutant copy—and vice versa. A feature of the gene-eviction procedure is that DNA sequences adjacent to the gene of interest are retrieved at the same time and so the method may be used for chromosome walking.

gene evolution: New genes in the genome evolve through gene duplication and diversification. At first, the function of the second copy of the gene is knocked out through a forbidden mutation and becomes a pseudogene. Then, with the help of new mutations, the duplicate gene assumes a new function.

gene expression. The process by which the biological information carried by a gene is released and made available to the cell, through transcription possibly followed by translation.

gene family: A group of genes that has arisen by duplication of an ancestral gene. The genes in the family may or may not have diverged from each other.

gene flow: The movement of genetic factors (genes) within a population or from one population to another, resulting from the dispersal of gametes (such as pollen by the wind) or zygotes (as by migration of fertile females, for example). (Birdsell, 1950)

gene frequency: The proportion of one allele of a pair or series present in the population or a sample thereof, that is, the number of loci at which a gene occurs, divided by the number of loci at which it could occur, expressed as a pure number between 0 and 1.

gene fusion: A DNA segment containing parts of different genes, e.g., the promotor of one gene and the coding region of another gene. *See* CHIMERIC GENE.

gene homology: The degree of similarity between the DNA sequences of genes.

gene interaction: The coordinated effect of two or more genes in producing a given phenotypic trait.

gene library: A large collection of cloning vectors containing a complete set of fragments of the genome of an organism.

gene machine: An idiomatic description of an automated oligonucleotide synthesizer.

gene manipulation: The formation of a new combination of heritable material by the insertion of nucleic acid molecules, produced by whatever means outside the cell, into any virus, bacterial plasmid, or any other vector system so as to allow their incorporation into a host organism in which they do not naturally occur, but in which they are capable of continued propagation.

gene map: A linear designation of mutant sites within a gene that is based upon the various frequencies of interallelic (intragenic) recombination.

gene mapping: Determination of the positions and relative distances of genes on chromosomes by means of their linkage.

gene mutation: These changes occur at or within a single gene; thus, they sometimes are called *point mutations.* As a result, a gene can mutate from one allelic form to another.

gene pair: The two copies of a particular type of gene present in a diploid cell (one in each chromosome set).

gene patenting: A sort of protection provided by a country's government to the discoverer of a new gene, genotype, genetic strain, gene test, or genetic procedure so that the detailed information can be declared publicly. Whereas scientists favor patenting of genes and of various human and life-forms, a large number of religious leaders oppose such patenting, thus leading to a burning debate. (Cole-Turner, 1995)

gene pool: Sum of all the different alleles in a random mating population at a particular period of time maintained through the mechanism of coadaptation. (Dobzhansky, 1951)

gene probe: A specific single-stranded DNA that is usually labeled with radioactive phosphorus to help the detection of a gene with a complementary sequence to which the probe can bind.

gene proper: That portion of the template strand containing coded information for a functional product.

gene reduplication. The duplication of genes in the chromosomes at cell division so that each daughter gene and chromosome are exact replicas of the originals.

gene regulation: Mechanism(s) such as repression or induction, positive or negative, that are responsible for controlling gene expression.

gene regulatory circuitries: The study of the evolution of genetic regulatory systems that are responsible for differences in gene expression at different developmental stages, in different tissues of the organism, and in different species.

gene regulatory mechanism in eukaryotes: Transcription of a producer gene occurs only if at least one of its receptors is activated by forming a sequence-specific complex with activator RNA. The activator RNA is synthesized by integrator genes in response to signals by sensor genes that are sensitive to external development signals.

gene regulatory mechanisms in prokaryotes: Basically, gene regulation involves interaction between RNA polymerase and promoter to initiate transcription, repressor and operator thus blocking passage of RNA polymerase, and repressor and inducer to inactivate the repressor. They are based on the presence or absence of a substrate (inducer), end product (corepressor), or product of a regulatory gene (repressor). Based on whether control is inducible or repressible, or whether control is positive or negative, several models of gene regulation in prokaryotes are conceived.

gene repression: The inhibition of gene expression. (Monod and Cohen-Bazire, 1953; Vogel, 1957)

gene rescue: A method allowing the reisolation of a transferred gene from stably transformed cell lines by way of its activity in *E. coli*. The recovery of plasmids is called *plasmid rescue.* (Perucho, Hanahan, and Wigler, 1980)

gene safety: In gene therapy, this term is used to emphasize that the new gene introduced into target cells of the patient should not harm the cells or, by extension, the organism.

gene scanning: A method in which the insertion of mutations at specific sites allows the determination of those DNA sequences needed for activity of a gene. A series of mutants are constructed in which the position of the block is systematically moved. In this way, a length of DNA can be scanned rapidly for regions of importance for the protein product of a gene or for regions involved in regulation of gene expression. (McKnight and Kingsburg, 1982)

gene sequencing: The technique of determining the order of the bases of a DNA molecule that constitute a gene.

gene shears: A genetically engineered molecule that can seek and destroy unwanted genetic messages within a cell. It may play a role in the prevention of plant and animal diseases, as well as in the destruction of viruses that invade human, plant, and animal cells.

gene splicing: The process of removal of introns and joining of exons from heterogenous nuclear RNA (hnRNA) to produce mature functional mRNA.

gene string: The central chromosome thread on which the genes are assumably arranged in sequence.

gene switch: A change in a cell or organism from the expression of one type of gene, or set of genes, to another related one, or set. Classic examples of gene switching occur during the development processes of higher organisms.

gene symbol: The symbol used to designate various genes, namely, dominant and recessive or wild and mutant genes.

gene synthesis: Chemical synthesis of a gene involving blocking of certain reactive groups of nucleotides and joining the same one by one. No template is used. For using this method, the nucleotide sequence of the gene to be synthesized must be known.

gene transfer: The passage of a gene or group of genes from a donor to a recipient organism.

genecology: *See* ECOLOGICAL GENETICS. Turesson (1923)

gene-environment interaction: A gene produces a particular effect only in the presence of a particular environment.

gene-for-gene hypothesis: This hypothesis states that for every gene for resistance in the host, there is a corresponding and specific gene for virulence in the pathogen. A disease fully develops only if the resistance gene in the host is overcome by a corresponding and specific virulence gene in the parasite. (Flor, 1956)

gene-gene interaction: Interaction between products of different genes leads to modification in phenotype or change in phenotypic ratios.

general resistance: Nonspecific host-plant resistance.

generalized recombination: Recombination between two double-stranded DNA molecules that share extensive nucleotide sequence similarity.

generalized transduction: Form of transduction in which any region of the host genome can be transduced. *See* SPECIALIZED TRANSDUCTION.

genetic advance: The expected gain in the mean of a population for a particular quantitative character by one generation of selection of a specified percent of the highest ranking plants.

genetic analysis: Deals with analysis of a gene in terms of its (a) nature, i.e., nuclear or cytoplasmic, dominant or recessive, autosomal or sex-linked (if applicable), (b) location on a particular chromosome, (c) position on the chromosome, and (d) nucleotide sequence.

genetic assimilation: Certain environmentally acquired characters could, through selection, become assimilated in the genotype. Genetic assimilation has experimental support. (Waddington, 1942)

genetic balance: The condition in which genetic components are adjusted in proportions that give satisfactory development. The term applies to individuals and populations.

genetic barrier: A physical or psychological condition that hinders or prevents the movement of individuals and thus of genetic information.

genetic base: Amount of genetic variability available for selection at a plant breeder's disposal.

genetic block: In biochemical genetics, a block in a step in the synthesis or segregation of a biochemical brought about by a mutant gene.

genetic cascade system: A system in which the completion of one event triggers the initiation of a second event. In genetics, usually a system for controlling the order in which genes are expressed.

genetic character: The expression of a gene as revealed in the phenotype.

genetic code: The rules that determine which triplet of nucleotides codes for which amino acid during translation.

genetic code dictionary: A table that gives information about the amino acids coded by different triplets of DNA or mRNA. *See* CODE DICTIONARY, also Table 1.

genetic code expansion: This deals with the incorporation of a nonstandard nucleotide to generate sixty-fifth genetic code so that ribosome-mediated incorporation of a nonstandard amino acid into a peptide is achieved. This generates more genetic variability.

genetic collapse: This occurs when the concentration of lethal genes in a population becomes great enough to cause the population to die out.

genetic death: Complete elimination of an allele from a population. Eliminating an allele is difficult. According to an estimate, it needs 30,000 generations for the genetic death of an allele. (Muller, 1950)

genetic decoding: *See* RECODING.

genetic deterioration: Refers to the fact that the genetic health of both plant and animal populations is deteriorating day by day because of our increased influence on natural processes through environmental pollution.

genetic disease: Any malfunctioning of the body directly caused by the malfunctioning of genes.

genetic dissection: The use of recombination and mutation to piece together the various components of a given biological function.

genetic distance (D): The average number of electrophoretically detectable codon substitutions per gene that have accumulated in the population studied since they diverged from a common ancestor.

genetic drift: Random change (in any direction) in gene frequencies due to sampling error. (Wright, 1921)

genetic engineering: The use of experimental techniques to produce DNA molecules containing new genes or new combinations of genes. This type of engineering deals with isolation or synthesis; and adding, removing, or replacing genes in order to achieve permanent and heritable changes in diverse forms of life bypassing all reproductive barriers. Also known as *recombinant DNA technology.*

genetic equilibrium: The condition of a population in which successive generations consist of the same genotypes with the same frequencies, with respect to particular genes or arrangements of genes. (Hardy, 1908)

genetic erosion: Loss of alleles (genes) due to different processes, such as breeding methods (e.g., inbreeding), advanced agricultural systems, genetic drift, etc.

genetic fine structure: The structure of the gene analyzed at the level of the smallest units of recombination and mutation (nucleotides). Benzer (1961) used *cis-trans* complementation test to study the fine structure of the gene.

genetic flexibility (resiliency): The inherent capacity of a genotype to vary on changes in conditions.

genetic homeostasis: The tendency of populations under selection to regress toward the original mean.

genetic identity (I): The proportion of genes whose products are not distinguishable by their electrophoretic behavior.

genetic induction: Stimulation of a specific gene by an inducer molecule. In bacteria operons, more than one gene may be induced simultaneously by the same inducer. Inducers may act directly or indirectly, usually through an effect on the RNA polymerase or on the promoter region upstream from the gene-coding sequence.

genetic information: The hereditary information contained in a sequence of nucleotide bases in chromosomal DNA or RNA that directs the development of a living example of that organism.

genetic instability: In some instances, genes mutate and back-mutate with a very high frequency. Certain controlling elements are considered to be responsible for this.

genetic isolation: Isolation of a variety or species by genetic means, e.g., cross-sterility in corn and cross-incompatibility of amphi-diploid species with parents or other species.

genetic load: The relative decrease in the mean fitness of a population due to the presence of genotypes (usually comprised of lethal genes) that have less than the highest fitness. (Muller, 1950)

genetic mapping: The process of determining the location of genes and the distance between genes on a chromosome.

genetic marker: A short DNA sequence, a whole gene, or a longer sequence of DNA acting as a landmark in the genome; also, any restriction fragment length polymorphism (RFLP) used to identify a specific linked gene or individual carrying a gene of interest.

genetic material: The chemical material of which genes are made, now known to be DNA in most organisms, RNA in a few.

genetic mutagen: A gene or a DNA sequence that causes other genes to mutate, e.g., the Ac-Ds (activator-dissociator) system in corn or the Dt gene in corn.

genetic perspective: The approaches used to study subject matter in genetics.

genetic polymorphism: The continued occurrence of two or more discontinuous genetic variants in a population in frequencies that cannot be accounted for by recurrent mutation. (Ford, 1940)

genetic risk assessment: Quantitative estimate of probable impact on the gene pool of subsequent generations from a specific mutagenic exposure.

genetic RNA: The RNA that acts as DNA or the genetic material.

genetic screening: The testing of individuals for a particular genetic trait.

genetic slippage: The loss of a potential genotype that may sometimes fail to reach an optimum level of phenotype due to genotype × environment interaction (hence, selected against).

genetic stock: A well-characterized strain, variety, or cultivar with known origin and inheritance used in genetic studies or breeding programs.

genetic switches: (1) Different control mechanisms act as switches that regulate different developmental phases of an organism. (2) Genetic devices that may turn transplant genes on and off at will. A tablet, for example, may turn on a transplated gene when required.

genetic system: All the factors, internal and external, that effect recombination in an organism. (Darlington, 1939b)

genetic toxicology: A subdiscipline of toxicology that identifies and analyzes the action of agents with toxicity directed toward the genetic material of the living systems.

genetic transformation: The radical transformation of an individual through transgenosis where a desired piece of DNA (gene) is nonsexually transferred within or between prokaryotes and eukaryotes following the recombinant DNA technology. The transferred gene replicates and expresses in its new host.

genetic uniformity: Homogeneity accruing from a narrow or single genetic base, e.g., a F_1 hybrid developed from a single male sterile line. This may lead to genetic vulnerability of the resultant crop, which may become highly prone to epidemics.

genetic variance: The phenotypic variance due to the presence of different genotypes in a population.

genetic vulnerability: *See* GENETIC UNIFORMITY.

genetically engineered male sterility: Induction of male sterility through recombinant DNA technology.

genetics: A branch of life sciences that deals with precise understanding of the nature, molecular structure, organization, biological function, alteration, regulation, and manipulation of the hereditary units, called genes, that are responsible for carrying out different life processes and for the transmission of biological properties from parents to offspring. Genetics also involves finding out ways and means to use genetic knowledge for improving the welfare of humankind. (Term originally coined by Bateson, 1905)

genic and plasmagenic interaction: Interaction between a plasmagene and a gene in the chromosome to produce an effect. For example, in paramecium, the plasma gene kappa and chromosomal dominant gene K interact to produce a killer phenotype.

genic disharmony: When genes contributed by two parents fail to work together in a hybrid or its progeny. This can be expressed at various stages of development and is associated with the process of nuclear metabolism, including DNA replication and transcription. Such individuals are at a selective disadvantage. Such disharmony serves as a postzygotic reproductive isolating mechanism.

genic equilibrium: The tendency of a particular gene frequency in a panmictic population to remain constant generation after generation.

genic explanation of hybrid vigor: Hybrid vigor is widely explained as the result of a dominance of linked genes. Many dominant genes are involved. Different inbreds contribute different dominant genes so that the F_1 has more than either parents.

genic male sterility: Male sterility controlled by nuclear genes.

genome: A haploid set of chromosomes of a diploid species is known as a genome. (Winkler, 1920)

genome analysis: Clarification of the homology between more related or less related species. This analysis is carried out by crossing and studying the chromosome pairing at meiosis in the F_1 hybrids.

genome cutter: A peptide nucleic acid (PNA) "clamp," consisting of two predesigned 8-bp sequences of PNA-linked together, can bind strongly and sequence specifically to DNA, hence known as bis-PNA, protecting the binding region from methylation. After removal of this bis-PNA, methylated DNA can be cut quantitatively at restriction sites selected by the PNA clamp. (Veselkov, Demidov, and Frank-Kamenetskil, 1996)

genome imprinting: A process that temporarily and erasably marks in different ways the genes passed on by females and males. Marked genes received by offspring from their mothers are consequently different from those received from their fathers.This imprinting involves differences in the expression of genes inherited from mother and father. (Surani, Barton, and Norris, 1984)

genome mapping: *See* GENE MAPPING.

genome mutation: An alteration involving one or more complete sets of chromosomes as in polyploidy (including autopolyploidy and allopolyploidy), or involving one or more approximately complete sets of chromosomes as in aneuploidy.

genome profiling: *See* GENOME IMPRINTING.

genomic: An adjective, usually applied to purified DNA from a particular source.

genomic allopolyploid: Polyploid species with two or more nonhomologous genomes. (Stebbins, 1947)

genomic clone: The term "genome" is taken to mean the entire complement of genetic information as it is found in the chromosomes. A genomic clone is one containing a fragment of chromosomal DNA rather than a copy of mRNA (i.e., cDNA).

genomic footprinting: A technique used to investigate areas of DNA-protein interaction. Cells in suspension are treated with dimethylsulfate

(DMS) or UV light to modify specific DNA bases that undergo a quantitative cleavage reaction. The modified sites are identified by genomic sequences. Proteins bound to DNA protect these areas from modification by DMS as subsequent chemical cleavage, leaving their "footprint" on the DNA. In contrast, UV light cleaves at sites of DNA-protein interaction. (Becker and Wang, 1984)

genomic *in situ* hybridization (GISH): This *in situ* hybridization technique involves the use of total genomic DNA of one species as a probe and total genomic DNA of another species as a blocking DNA. GISH is exclusively based on fluorescence *in situ* hybridization (FISH). This method is useful for identification of interspecific/intergenomic hybridization and addition/translocation lines.

genomic library: A collection of clones sufficient in number to include all the genes of a particular organism. Also known as *gene library.*

genosome: The physical part of a chromosome at the point where a gene is located. (Ravin, 1963)

genospecies: A group, all the members of which are genotypically identical.

genotoxic: Those agents that produce alterations in the nucleic acids at subtoxic levels resulting in modified hereditary characteristics or DNA inactivations.

genotype: A description of the genetic composition of genes, a cell, or an organism. The genotype is relatively stable throughout the life of an organism. (Johannsen, 1909)

genotype frequency: The relative proportion of a particular genotype in a population.

genotype-environment correlation: The nonrandom assortment of particular genotypes among particular environments within a given population.

genotype-environment interaction: The interplay of a specific genotype and a specific environment that affects the phenotype. The extent and nature of this interaction varies with each genotype and environment. The variance due to this interaction (V_I) is part of the total phenotypic variance of a trait.

genotypic control: The control of chromosome behavior by the genotype.

genotypic environment: The aggregate of all the genes considered as acting on one or more of the genes.

genotypic ratio: The ratio of different genotypes in a segregating generation. Genotypic ratio of F_2 of a monohybrid cross is 1AA: 2Aa:1aa.

genotypic value: The measured value of the contribution of the genotype.

genotypic variation: Two or more individuals developing in the same environment and coming to possess different phenotypes indicates that these individuals have different genotypes. Genotypic variation arises due to changes in the germ plasm.

genuine pleiotropism: The controlling, by a single gene, of two or more different effects which it produces directly and by the use of different mechanisms. Spurious pleiotropism may be brought about in two ways. Either the gene does two (or more) things directly, but by means of the same mechanism, or else it has one effect which in turn causes other things to happen.

geographic speciation: The formation of a new species during a period of geographical isolation. Any type of geographical feature, whether mountain, river, or other type of terrain, that prevents gene flow between populations is known as *geographic barrier.* A distinctive population whose distribution is correlated to a geographical region. A population that results from broad, climate-selective forces is known as a *geographical subspecies.*

germ cell: Generative cells in a multicellular organism that are distinct from the somatic cells. (Engler and Prantl, 1897)

germ line: The line of cells that produces gametes.

germinal mutation: Mutation in a tissue that ultimately forms sex cells which may be passed on to the next generation.

germplasm: (1) The genetic makeup of an organism—the sum total of its genes, both dominant and recessive. (2) The potential hereditary materials within a species, taken collectively. (Weismann, 1883, 1885)

germplasm collection: A collection of genotypes of a particular species, from different sources and geographic locations, used as source materials in plant breeding. *See* WORLD COLLECTION.

germplasm theory: A theory stating that multicellular organisms give rise to two types of tissues, somatoplasm and germ plasm. Somatoplasm is essential for the functioning of the organism but lacks the property of entering into sexual reproduction. Changes in somatic tissues are not heritable. Germ plasm is set aside for reproductive purposes. A change in germ plasm could be heritable.

glucose-sensitive operons: Those operons whose functioning is blocked by the presence of glucose. Glucose indirectly lowers the cycle AMP level thereby blocking a necessary positive control signal.

Goldberg-Hogness box: *See* TATA BOX. (Goldberg, 1979)

gones: The four cells, or their nuclei, that are normally the immediate product of meiosis.

gradual speciation: True speciation that occurs through changes in populations.

graft hybrid: A hybrid-like product obtained after grafting between different species or varieties. In most cases, the graft hybrid is a chimera composed of different cell layers from the parents.

Gram-negative bacteria: One of the two categories into which all bacteria are divided according to their Gram's staining reaction. Bacteria are first stained with crystal violet and then with iodine. Treatment with acetone will destain Gram-negative bacteria. Such bacteria have a complex cell wall in which an outer membrane overlays the rigid peptidoglycan matrix. Enteric bacteria such as *Escherichia coli* and *Salmonella typhimurium* are Gram-negative. The staining method is named after its inventor, Christian Gram, the nineteenth-century Danish bacteriologist.

Gram-positive bacteria: Bacteria that resist decolorization by acetone during the Gram's staining reaction. Gram-positive bacteria have a simpler cell wall than do Gram-negative bacteria. It is made predominantly of peptidoglycan and is not overlaid by an outer membrane. *Bacillus subtillis* and *streptomyces* are genetically important Gram-negative bacteria. *See* GRAM-NEGATIVE BACTERIA.

graphical genotype: The portrait of restriction fragment polymorphism data in a graphical form for description of an entire genome in a single graphic image. It describes the parental origin and allelic composition throughout the entire genome. (Young and Tanksley, 1989)

gratuitous inducers: These resemble authentic inducers of transcription but are not substrates for the induced enzymes.

gRNA: *See* GUIDE RNA.

group I introns: Self-splicing introns that do not require an external nucleotide for splicing. The intron is released in a lariat form.

group II introns: Selection for traits that would be beneficial to a population at the expense of the individual possessing the trait.

group selection: Process in which natural selection acts upon a few individuals in a group to select characteristics that are beneficial to the group rather than the individual. (Wynne-Edwards, 1962)

group-transfer reaction: Reaction (excluding oxidations or reductions) in which molecules exchange functional groups.

GT-AG rule: Refers to the fact that with introns present in nuclear protein-coding genes, the first two nucleotides of the intron are GT and the last two are AG.

GTP (guanosine triphosphate): An energy-rich molecule required for the formation of peptide bonds between amino acids.

guanine: 2-amino, 6-oxypurine, one of the nucleotides found in DNA and RNA.

guide RNA: These are short RNA molecules that can form perfect hybrids with mRNA to be edited, and they possess nucleotide sequences at their 5′ ends that are complementary to the sequences of the mRNAs immediately downstream of the pre-edited regions (PER). Guide RNAs are of ~ 60 nucleotide length. (Church, Stonimski, and Gilbert, 1979; Blum et al., 1991)

guide sequence: An RNA molecule (or a part of one) that hybridizes to eukaryotic mRNA and aids in the splicing of intron sequences. Guide sequences may be either external (EGS) or internal (IGS) to the mRNA being processed and may hybridize to either intron or exon sequences close to the splice junction.

gyrase: An enzyme that relaxes supercoiling caused by unwinding of a double helix of DNA. (Gellert et al., 1976)

H DNA: *See* DNA HELIX H. (Hentschel, 1982)

Hae III: A type II restriction enzyme from the bacterium *Haemophilus aegyptius* that recognizes the DNA sequence shown below and cuts at the site indicated by the arrows. Hairpin loops are formed by palindromic sequences.

```
            ↓
5′  G  G  C  C  3′
3′  C  C  G  G  5′
            ↑
```

hairpin loop: A double helical region formed by base pairing between adjacent (inverted) complementary sequences in a single strand of RNA or DNA.

half chromatid: One of the two strands in a chromatid.

half-chromatid conversion: A type of gene conversion that is inferred from the existence of nonidentical sister spores in a fungal octad showing a non-Mendelian allele ratio.

haplodiploidy: The sex-determining mechanism found in some insect groups among which males are haploid and females are diploid.

haploid: Refers to a cell that contains a single copy of each chromosome. (Strasburger, 1905)

haploid number: The number of chromosomes found in a normal gamete.

haploidization: Production of a haploid from a diploid by progressive chromosome loss.

haplontic selection: Selection of haploid cells under irradiation. The diplontic sieve is more rigorous than the haplontic sieve.

haplophase: That part of the life cycle in which the gametic chromosome number is found in reproductive cells.

haplosis: Establishment of the gametic chromosome number, usually by meiosis.

haplotype: A combination of very closely linked alleles or markers that tend to be transmitted as a unit to the next generation; also refers to a pattern of DNA restriction fragments for a chromosome region that can be recognized in an autoradiograph.

haplo-viable mutants: These prevent normal functioning of pollen but are transmitted through the egg carrying them. Also called *male gamete eliminators*.

Hardy-Weinberg equilibrium: Occurrence in a population of three genotypes, two homozygous and one heterozygous (e.g., AA, Aa, and aa), where the frequencies of the three genotypes are p^2 for the homozygous dominants, 2 pq for the heterozygotes, and q^2 for the homozygous recessive, in which p equals frequency of A and q equals frequency of a. At these frequencies, no change in genotypic frequencies occurs as long as isolation mechanisms, selection mutation frequencies, and drift are constant. (Hardy, 1908; Weinberg, 1908)

harlequin chromosomes: Sister chromatids that stain differently so that one appears dark and the other light in color.

harvest index (HI): The ratio of economic yield to biological yield (HI = Yeco/Ybiol); high HI reveals rapid translocation of photosynthate from efficient source to powerful sink.

HAT selective medium: A medium containing hypoxanthine, aminopterin, and thymidine that is used to select somatic hybrid cells.

heat mutagenesis: Induction of mutations at above melting temperature by strand separation, depurination, and chain breakage.

helicase: The enzyme responsible, during DNA replication, for breaking the hydrogen bonds that hold the double helix together.

helix: A structure with spiral shape, e.g., the double helix model of DNA proposed by Watson and Crick in 1953.

helix-destabilizing proteins: Proteins that bind the single-stranded portion of the DNA molecule, preventing it from assuming a double-stranded conformation during replication. (Alberts et al., 1989)

helix-turn-helix motif: Configuration found in DNA-binding proteins consisting of a recognition helix and a stabilizing helix separated by a short turn.

hemizygote: A diploid individual that has lost its copy of a particular gene (for example, because a chromosome has been lost) and therefore has only a single copy.

hereditary univalents: Chromosomes that have evolved a system of approximately even distribution in segregation without going through the process of pachytene pairing. *Compare* SPONTANEOUS UNIVALENT.

heredity: The passage of characteristics from parents to offspring.

heritability (H^2): A statistic measuring degree to which the total phenotypic variation is a result of genetic factors. Also known as *true heritability.* (Lush, 1941)

heritability in broad sense (H_b): An estimate of heritability measuring contribution of all genetic components contributing toward the phenotype.

heritability in narrow sense (H_n): An estimate of heritability measuring contribution of only additive genetic effects contributing toward the phenotype.

hermaphroditism: Reproductive organs of both sexes present in the same individual or in the same flower in higher plants.

heteroallele: The allele that arises due to base pair replacement in different mutons; those alleles between which recombination is theoretically possible. (Roman, 1956)

heterobrachial aberration: Chromosomal aberration involving both arms of a chromosome.

heterobrachial duplication: When a duplicate segment is present in the other arm of the chromosome.

heterocaryosis: The presence of two or more genetically different nuclei within the single cell of a mycelium.

heterochromatic: Entire chromosomes or portions of the chromosomes that do not manifest the usual prophase-telophase transformations and appear to lack genes with major phenotypic effects. (Heitz, 1928, 1929)

heterochromatin: The regions of a chromosome that appear relatively condensed and stain deeply with DNA-specific stains. These are composed of 250 Å chromatin *(solenoid chromatin).*

heterodimer: A dimeric protein having two different polypeptides.

heteroduplex DNA: A base-paired structure formed between two deoxypolynucleotides that are not entirely complementary because they are derived from different duplex DNA molecules.

heteroduplex DNA model: A model that explains both crossing-over and gene conversion by assuming the production of a short stretch of heteroduplex DNA (formed from both parental DNAs) in the vicinity of a chiasma.

heteroduplex mapping: The use of heteroduplex analysis to determine the location of various inserts, deletions, or heterogeneities in a piece of DNA. (Davis and Hyman, 1971)

heteroduplex modified DNA: DNA having no methyl group on a particular adenine or cytosine on only one strand included in a host specificity site.

heterogamic mating: Mating between individuals of different genotypes.

heterogamy: Matings between unlike individuals; sometimes the preferential mating of individuals of dissimilar genotypes of phenotypes.

heterogeneity: Lack of agreement among different bodies of data in regard to the value of one or more parameters of which they are capable of yielding estimates.

heterogeneous nuclear RNA (hnRNA): The nuclear RNA fraction that comprises the unprocessed transcripts synthesized by RNA polymerase II. (Scherrer and Darnell, 1962)

heterogenic: A population or a gamete containing more than one allele of a particular gene or genes. *Compare* HOMOGENIC.

heterokaryon: A cell possessing two or more genetically different nuclei in a common cytoplasm, usually derived, as a result of fusion of protoplasts belonging to diverse karyons (origin), through somatic hybridization.

heterologous duplication: When a duplicate segment is present in the nonhomologous chromosome.

heterologous probing: The use of a labeled nucleic acid molecule to identify related molecules by hybridization probing.

heteromorphic bivalent: When two chromosomes of the bivalent differ structurally due to chromosomal aberration. (Carothers, 1917)

heteromorphic chromosomes: Chromosomes of which the members of a homologous pair are not morphologically identical. (Winkler, 1916)

heteromultimeric proteins: A protein that consists of nonidentical subunits, which are coded by different genes.

heteroploid: The term given to a cell culture when the cells comprising the culture possess nuclei containing chromosome numbers other than the diploid number. This is a term used only to describe a culture and is not used to describe individual cells. Thus, a heteroploid culture would be one that contains aneuploid cells.

heteropolymer: An artificial nucleic acid molecule comprised of a mixture of different nucleotides.

heteropyknosis: A condition in which chromosomes or chromosome regions are out of phase in respect to staining properties and the coiling cycle.

heterosis: (1) The increased vigor, growth, size, yield, or function of a hybrid progeny over the parents that results from crossing genetically unlike organisms. (2) The increase in vigor or growth of a hybrid progeny in relation to the average of the parents. (Shull, 1911)

heterosomal: A chromosomal aberration in which one of the homologs of a chromosome pair is involved.

heterozygosity index: A measure of the genetic variance in a population, with respect to one locus, stated as the frequency of heterozygotes for that locus.

heterozygote: An individual having a heterozygous gene pair.

heterozygote advantage: A selection model in which heterozygotes have the highest fitness.

heterozygous: A gene pair having nonidentical alleles in two homologous chromosomes (e.g., Aa is a heterozygous genotype). (Bateson and Saunders, 1902)

hexaploid: Usually, a diploid with duplicate sets of three different genomes. Common bread wheats (42 chromosomes) are hexaploids in contrast to primitive wheat, which has 14 chromosomes (a diploid).

HFR strain: A mating strain of *E. coli* that carries the fertility factor, F, integrated into the chromosome and that causes a high frequency of recombination following conjugation with an F- cell.

HGPRT (hypoxanthine phosphoribosyl transferase): Enzyme that catalyses transfer of the phosphoribosyl moiety to hypoxanthine and guanine. *See* HAT SELECTIVE MEDIUM.

Hha restriction enzyme: Type II restriction enzyme isolated from *Haemophilus haemolyticus* that recognizes the tetranucleotide target sequence. Cleavage of the DNA duplex occurs at the positions arrowed to yield a fragment of DNA with protruding cohesive ends that are two bases in length.

```
           ↓
5′  G  C  G  C  3′
3′  C  G  C  G  5′
        ↑
```

high frequency recombination strain: *See* HFR STRAIN.

highly repetitive DNA: The first component of reassociate; equated with satellite DNA.

highly repetitive sequences: When a number of copies of a sequence in eukaryotic DNA is 10^6 to 10^7. These repeats are usually found in centromeric regions.

high-mobility group (HMG) proteins: nonhistone proteins (e.g., HMG-14 and HMG-17) that are involved in the switch from an inactive to an active chromatin structure by associating themselves with nucleosomes. (Ding et al., 1994)

high-performance liquid chromatography (HPLC): Chromatography carried out in a vacuum, in which the solvent phase is forced through the solid phase under high pressure, improving speed, and under certain circumstances, resolution.

high-performance ultrafiltration (HPUF): A type of high performance liquid chromatography that uses a gel filtration column.

high-voltage electron microscopy: Type of transmission electron microscopy that accelerates electrons through 10^6 volts and can be used for examining sections as thick as 1 μm.

hinge DNA: *See* DNA HELIX H.

histone: One of the basic proteins that makes nucleosomes and has a fundamental role in chromosome structure. Five major species of histones (H1, H2A, H2B, H3, and H4) are present in chromatin in a molar ratio of 0.5:1:1:1:1. Genes coding for histones lack histones and polyadenylation of mRNA. Presence of a 14 bp G+C-rich palindromic structure is followed by A-C-C-A near the 3′ ends of their mRNAs. (Kossel, 1884)

histone-dependent activation: This is the first stage in the activation of genes in which activator proteins, at the upstream activator sequence (UAS), directly or indirectly cause the histone core particles to dissociate from *TATA box* and form a preinitiation complex and thereby generate a basal level of transcription.

histone-independent activation: This is the second stage in the activation of genes in which activators stimulate the preinitiation complex to maximize production of mRNA.

histones as gene regulators: Histones are not only important basic proteins that impart structure to eukaryotic chromosomes but they also play important roles in the cell cycle by undergoing reversible modifications, such as acetylation of lysine, phosphorylation of serine and threonine, and ubiquitnation of lysine.

hnRNA: Heterogeneous nuclear RNA, RNA polymerase II transcript, and their processed intermediates. *See* HETEROGENEOUS NUCLEAR RNA.

Hogness-Goldberg box: A nucleotide sequence that makes up part up of the eukaryotic promoter. *See* TATA BOX.

Holliday structure: An intermediate structure believed to be formed during recombination between two DNA molecules. (Sigal and Alberts, 1972)

holoenzyme: The associated form of a multisubunit enzyme with dissociable subunits. *See* CORE ENZYME.

holorepressor: A functional repressor protein consisting of an aporepressor plus corepressor complex. *See* APOREPRESSOR and COREPRESSOR.

homeo domain protein: The sixty-amino acid protein translated from the homeobox. It consists of three main regions—(1) variable region, which regulates the protein's specific activity, (2) hinge region, a small connective region, and (3) homeo domain, which binds homeoprotein to the DNA or RNA.

homeobox: A highly conserved segment of DNA found in all homeotic genes. This sequence may code for the part of a regulatory protein that binds to DNA. (McGinnis et al., 1984)

homeogenetic induction: Process by which a differentiated cell can induce differentiation in an adjacent cell not previously so committed. The term is used chiefly in relation to plant cells.

homeologous: Having partial homology.

homeologous pairing: Partial pairing between chromosomes due to their partial homology.

homeosis: Ability of a genotype to produce the same phenotype under varied environments. (Bateson, 1894)

homeostasis: The totality of the steady state maintained in an organism through the coordination of its physiological processes—an autoregulation device conferring genetic resiliency or buffering capacity to the organism. (Cannon, 1929)

homeotic genes: The genes defined by mutations that convert one body part into another. (Goldschmidt, 1945)

homoallele: The alleles that arise due to base pair replacement in the same muton; recombination is theoretically not possible between homoalleles.

homoallelic: The genes similar to each other in function but located in homeologous chromosomes with different evolutionary origins as in the case with polyploids of duplicated, triplicated, etc. genes. (Washington, 1971)

homobrachial duplication: When a duplicate segment is present in the same arm of the chromosome.

homodimer: A dimeric protein having two identical polypeptides.

homoduplex modified DNA: DNA having a methyl group on a particular adenine or cytosine of both strands included in a host specificity site.

homoduplex unmodified DNA: DNA having no methyl group on a particular adenine or cytosine on both strands included in a host specificity site.

homoeosis: The alternation of one organ of a segmental or homologous series from its own characteristic forms to that of some other member of the series.

homogamic mating: Mating between individuals of the same genotype.

homogamy: Mating between individuals of like phenotype, genotype, or structure.

homogeneous population: Uniform—all of a single type.

homogenic: A population or a gamete containing only one allelomorph of a particular gene or genes. *Compare* HETEROGENIC. (Fisher, 1928)

homokaryon: A cell possessing two or more genetically identical nuclei in a common cytoplasm, derived as a result of cell to cell fusion.

homolog: One of a pair of homologous chromosomes.

homologous chromosomes: Two or more identical chromosomes (one received from the father and the other received from the mother) that are able to undergo synapsis during prophase of meiosis I.

homologous genes. Two genes from different organisms, and therefore of different sequence, that code for the same gene product.

homologous recombination: Recombination between two double-stranded DNA molecules that share extensive nucleotide sequence similarity.

homomorphic bivalent: When two pairing partners are structurally similar.

homomultimeric proteins: Proteins that consist of identical subunits.

homopolymer: An artificial, nucleic acid molecule comprising just one nucleotide.

homopolymer tail: A sequence of only one sort of nucleotides at the end of a DNA or RNA molecule. Most eukaryotic mRNAs have a homopolymer tail of pA at the 3′ end.

homosequential species: A group of related species having identical gene arrangements in all their chromosomes, as indicated by their banding patterns. (Blakeslee, 1904)

homozygote: An individual having a homozygous gene pair. Homozygotes produce only one kind of gamete with respect to a particular locus and therefore "breed true."

homozygous: A gene pair having identical alleles in two homologous chromosomes (e.g., AA and aa are homozygous genotypes).

honeycomb method: To minimize adverse effects of interplant competition, plants may be grown on a hill surrounded by six other hills which form a hexagon, thus sharing a common microenvironment (one hexagon encompasses seven hills).

Hoogsteen base pairing: Type of base pairing that occurs in triplex stranded DNA (H DNA) and RNA structures. Hoogsteen base pairing is seen as T=A=T and C=G≡C triple base pairing.

horizontal resistance: The type of resistance that is governed by "minor" genes or polygenes which act additively together to control resistance which is race nonspecific in nature and does not reveal a gene-for-gene hypothesis. This type of resistance is relatively difficult to identify.

horizontal transmission: Transfer of genetical material by infection-like processes, including transduction. *Compare* VERTICAL TRANSMISSION.

host: An organism able to support the replication of a plasmid or virus.

host-cell reactivation: The repair of the damaged DNA of viruses by host-cell repair mechanisms.

host-controlled modification: After DNA replication, the host methylates its own DNA with the help of its methylase enzyme. This enzyme is composed of polypeptide products of hsdM and hsdS genes in the ratio 2:1.

host-controlled restriction: DNA synthesized in one cytoplasm that undergoes cleavage when it enters the host cytoplasm. The restriction endonuclease is made up of three different polypeptides, in the ratio 2:2:1, synthesized by three genes, hsdR, hsdM, and hsdS, respectively.

host-parasite specificity: Refers to the ability of a pathogen to pathogenize distinct groups of plants. It is known to operate at genus, species, and genotypic level.

hot spot: A site where the frequency of mutation (or recombination) is very much increased. (Benzer, 1955)

housekeeping genes: The genes whose products are required by all types of cells at all times. Their activity is controlled by constitutive factors. *See* CONSTITUTIVE GENE.

Hpa I: A type II restriction enzyme from the bacterium *Haemophillus parainfluenzea* that recognizes the DNA sequences shown below and cuts at the sites indicated by the arrows:

```
            ↓
5′  G  T  T  A  A  C  3′
3′  C  A  A  T  T  G  5′
            ↑
```

Hsa I: A type II restriction enzyme from the bacterium *Haemophillus haemolyticus* that recognizes the DNA sequences shown and cuts at the sites indicated by the arrows:

```
              ↓
5′  G  C  G  C  3′
3′  C  G  C  G  5′
       ↑
```

hybrid: An individual that results from a cross between two genetically unlike parents.

hybrid breakdown: Production of weak or sterile F_2 progeny by vigorous fertile F_1 hybrids.

hybrid cell: The term used to describe the mononucleate cell that results from the fusion of two different cells and leads to the formation of synkaryon.

hybrid DNA: DNA whose two strands have different origins.

hybrid gene: A gene composed of a fragment of the gene of interest ligated to some other heterologous DNA sequence that permits regions to respond to regulatory signals.

hybrid inferiority: A hybrid is inferior to either of the parents.

hybrid inviability: Hybrids either do not survive or they do not reach sexual maturity.

hybrid plasmid: A plasmid that contains an inserted piece of foreign DNA.

hybrid resistance. The rejection of parental tissue grafts by the F_1 hybrids of two inbred strains.

hybrid screening: A radioisotope technique used to determine whether a hybrid plasmid contains a particular gene or DNA region.

hybrid selection: The process of choosing plants possessing desired characteristics from among a hybrid population; identification and isolation of those desirable for further studies, propagation, etc.

hybrid sorting: The selection of superior plants among haploids derived from F_1 hybrids through anther culture.

hybrid sterility: Hybrid individuals are sterile.

hybrid vehicle: An episome or plasmid containing an inserted piece of foreign DNA. Also known as *vector.*

hybrid vigor: Superiority of a hybrid over the better parent in one or more traits. This is also known as *positive heterosis*.

hybrid-arrested translation (HART): A method used to identify the polypeptide coded by a cloned gene. The method involves DNA-RNA hybridization and is as follows: (a) A crude cellular mRNA preparation, composed of many types of mRNA, is hybridized with a cloned, denatured DNA; (b) mRNA homologous to the cloned DNA will anneal to it; then (c) the rest of the mRNA molecules are put into an *in vitro* translation system, and the protein products are compared with the proteins obtained by use of the whole mRNA preparation. (Izant and Weintraub, 1985)

hybridization: The pairing of complementary RNA and DNA strands to produce an RNA-DNA hybrid.

hybridization fingerprinting: A technique used to identify clone overlaps. It is designed to combine the high data rates achievable by hybridization techniques and the favorable scaling behavior and the repeat intensity of the gel fingerprinting approaches. In this approach, filters carrying a large number of DNA spots are generated by colony lysis. These DNA spots are hybridized to oligonucleotide probes or more complex probe mixtures complementary to many positions in the genome. (Sternberg, 1990)

hybridization *in situ*: Finding the location of a gene by adding specific radioactive probes for the gene and detecting the location of the radioactivity on the chromosome after hybridization.

hybridization kinetics: The speed with which a probe binds with complementary strands. It gives an estimate of the number of copies of a sequence present.

hybridization probing: A method that uses a labeled nucleic acid molecule to identify complementary or homologous molecules through the formation of stable base-paired hybrids.

hybridize: (1) To form a hybrid by performing a cross; (2) to anneal nucleic acid strands from different sources.

hybrid-released translation (HRT): A method used to identify the polypeptide coded by a cloned gene.

hydroxyapatite: A form of calcium phosphate that binds double-stranded DNA.

hyperchromacity: The increase in optical density that occurs when DNA is denatured.

hypermorph: An allele having an effect similar to but greater than that of the wild form. (Muller, 1932)

hyperploid: An aneuploid with a small number of extra chromosomes missing. (Bĕlăr, 1928)

hypersensitive site for DNase I: A short region of chromatin detected by its extreme sensitivity to cleavage by DNase I and other nucleases; probably comprises an area from which nucleosomes are excluded.

hypersensitivity: Premature death of the infected host tissues, as well as inactivation and localization of the pathogen, thus restricting the spread of obligate parasite.

hypomorph: An allele having an effect similar to but less than that of the wild form. (Muller, 1932)

hypoploid: An aneuploid with a small number of chromosomes missing. (Bĕlăr, 1928)

hypostatic gene: The gene whose expression is suppressed by other genes. (Bateson and Punnett, 1906)

I_1, I_2, I_3, etc: Symbols used to designate first, second, third, etc. inbred generations. *See* S_1, S_2. . . .

identical by descent: True of two genes that are identical in nucleotide sequences because they are both derived from a common ancestor. (Malecot, 1948)

identical in structure: True of two genes that are identical in nucleotide sequences regardless of whether or not they are both derived from a common ancestor.

ideochromatin: Nuclear chromatin controlling cell division but otherwise dormant.

ideotype: A physiologically efficient, model plant type—a biological model of a plant frame that can exploit the available resources (solar energy, nutrients, etc.) most efficiently to produce maximum biomass, hence, maximum economic yield.

idiogram: A diagrammatic representation of the karyotype of a species or variety based on total chromosome length, arm length ratio, centromere position, and nucleolus organizer position. (Navashin, 1912)

idling reaction: The production of guanosine tetraphosphate (3′-ppGpp-5′) by the stringent factor when a ribosome encounters an uncharged tRNA in the A site.

ignorant DNA: *See* SELFISH DNA.

illegitimate recombination: Recombination between two double-stranded DNA molecules that have few orders of structure among many biomolecules.

immunity in phages: The ability of a prophage to prevent another phage of the same type from infecting a cell. It results usually from the synthesis of phage repressor by the prophage genome.

immunity in plasmids: The ability of a plasmid to prevent another of the same type from becoming established in a cell. It results from interference with the ability to replicate.

immunity in transposons. The ability of certain transposons to prevent others of the same type from transposing to the same DNA molecule.

immunoscreening: The use of an antibody to detect a polypeptide synthesized by a cloned gene.

***in situ* hybridization:** Hybridization performed by denaturing the DNA of cells squashed on a microscope slide so that reaction is possible with an added single-stranded RNA or DNA; the added preparation is radioactively labeled and its hybridization is followed by autoradiography.

***in situ* nucleic acid hybridization:** The annealing of radioactive, single-stranded DNA or RNA probes to denatured cellular DNA on microscope slides and their detection by autoradiography. This method cytologically permits the localization of DNA sequences complementary to the probes in intact chromosomes. (Gall and Pardue, 1969; John, Birnsteil, and Jones, 1969)

***in vitro* complementation:** Complementation of allelic sequences demonstrated in a test tube.

***in vitro* marker:** A mutation that allows identification *in vitro* of a cell line possessing the marker. Such mutations are commonly expressed in culture as virus or drug resistance mutations. *Compare* *IN VIVO* MARKER.

***in vitro* mutagenesis:** Any one of several techniques used to produce a specified mutation at a predetermined position in a DNA molecule.

***in vitro* packaging:** The synthesis of infective λ particles from a preparation of λ capsid proteins and a concatamer of DNA molecules separated by cos sites.

***in vitro* pollination:** Pollination performed aseptically *in vitro*. Pollen is applied directly to ovules in an attempt to overcome various types of prezygotic incompatibilty that otherwise inhibit fertilization.

***in vitro* propagation:** Propagation of plants in a controlled, artificial environment, using plastic or glass culture vessels, aseptic techniques, and a defined growing medium.

***in vitro* protein synthesis:** The incorporation of amino acids into polypeptide chains in a cell-free system.

***in vitro* transcription (cell-free transcription):** The specific and accurate synthesis of RNA in the test tube with purified DNA preparations as a template. So-called "coupled system" may be obtained from *E. coli,* which carry out both mRNA synthesis and its translation into protein. For eukaryotes, separate cell-free systems have to be set up to demonstrate the activity of the three functionally distinct RNA polymerase complexes.

***in vitro* transformation:** A heritable change, occurring in cells in culture, either intrinsically or from treatment with chemical carcinogens, oncogenic viruses, irradiation, transfection with oncogenes, etc., and leading to the acquisition of altered morphological, antigenic, neoplastic, proliferative, or other properties.

***in vitro* translation (cell-free translation):** The synthesis of proteins in a test tube from purified mRNA molecules using cell extracts containing ribosomal subunits, the necessary protein factors, tRNA molecules, and aminoacyl tRNA synthetases. ATP, GTP, amino acids, and an enzyme system for regenerating the nucleoside triphosphates are added to the mix. Prokaryotic translation systems are usually prepared from *E. coli* or the thermophilic bacterium *Bacillus stearothermophilus*. Eukaryotic systems usually employ rabbit reticulocyte lysates or wheat germ.

***in vivo* marker:** A mutation that occurs in the organism but also allows detection in tissue culture—for example, genes causing galactosemia being detectable both in the phenotype of an individual and *in vitro* in tissue cells derived from such individuals. *Compare* *IN VITRO* MARKER.

inactivating DNA alteration: Any DNA damage that blocks DNA replication. If not eliminated by DNA repair processes, such damage is usually lethal to the cell. Occasionally, it may give rise to gene mutations and frequently, by mispairing to chromosome structural changes. (Freese, Phaese, and Freese, 1969)

inbred line: (1) A pure line usually originating by self-pollination and selection; (2) the product of inbreeding.

inbred-variety cross: The F_1 cross of an inbred line with a variety.

inbreeding: The mating of genetically related individuals.

inbreeding coefficient (F): The degree of relationship between parents and progeny indicating intensity of inbreeding. In other words, the probability that the two alleles are related by descent. A quantified degree of inbreeding. (Wright, 1922)

inbreeding depression: The loss of vigor and physiological efficiency of an organism on inbreeding, leading to reduction in size and fecundity, etc. Expressed in filial generations, it is $(F_2 - F_1)/F_1$, and the inbreeding effect is represented as its percentage (i.e., $100\ (F_2 - F_1)/F_1$).

inbreeding minimum: The stage reached after a long period of inbreeding. Because of the homozygosity attained, vigor is not further decreased but remains constant if inbreeding is continued.

incipient species: Populations that are too distinct to be considered as subspecies of the same species but not sufficiently differentiated to be regarded as different species. Also called *semispecies*.

incompatibility: The inability of certain types of plasmids to coexist in the same cell. (Stout, 1918)

incompatibility group: A group comprised of different plasmids that are unable to coexist in the same cell.

incomplete digest: *See* PARTIAL DIGEST.

incomplete dominance: The situation where neither of a pair of alleles displays dominance, and the phenotype of the heterozygote is intermediate between the phenotypes of the two alternative homozygotes.

incomplete linkage: Linked genes that reassociate through genetic crossing-over.

incomplete penetrance: The extent to which a phenotype is expressed is less than 100 percent when measured as a proportion of individuals with a particular genotype that actually display the associated phenotype.

incomplete resistance: A type of resistance that is not complete and that causes slow rusting. *See* SLOW RUSTING.

independent characters: Characters governed by genes showing independent assortment.

independent genes: Genes that show independent assortment.

independent segregation (assortment), principle of: Different segregating gene pairs behave independently during meiosis. *See* MENDEL'S LAWS.

indeterminate: Descriptive of an inflorescence in which the terminal flower is last to open. The flowers arise from axillary buds, and the floral axis may be indefinitely prolonged by a terminal bud.

index selection. *See* TANDEM SELECTION.

indirect end labeling. A technique for examining the organization of DNA by making a cut at a specific site and isolating all fragments containing the sequence adjacent to one side of the cut; it reveals the distance from the cut to the next break(s) in DNA.

individual buffering: A mechanism promoting genotypic adaptation of homogeneous populations (e.g., pure-line or F_1 hybrids) that tend to stabilize their production under different environments. It operates at allelic level within an individual. Also known as *developmental homeostasis*.

induced chromosome break: Chromosome breaks caused by some agent usually external to the chromosome, such as radiation or chemicals. It can be produced genetically (e.g., the Ac-Ds system in maize).

induced mutations: Genetic changes produced by some physical or chemical agent or under changed growth conditions.

inducer: A molecule that induces expression of a gene or operon by binding to a repressor protein and thereby preventing the repressor from attaching to its operator site. Effectors in inducible operons are known as inducers. (Jacob and Monod, 1961b)

inducible control: In this case, a substrate acts to induce the production of the enzymes. Repression occurs in the absence of the substrate.

inducible enzyme: An enzyme synthesized only in the presence of an effector. *See* ADAPTIVE ENZYME.

inducible system: A system (a coordinated group of enzymes involved in a catabolic pathway) is inducible if the metabolite upon which it works causes transcription of the genes controlling these enzymes. These systems are primarily prokaryotic operons.

induction: (1) The switching on of gene expression; (2) initiation of a structure, organ, or process *in vitro*.

inert gene: A gene that, through apparent lack of mutation or effect in disturbing balance, has been assumed to be inactive. Often associated with heterochromatin.

inflorescence: (1) A flower cluster; (2) the arrangement and mode of development of the flowers on a floral axis.

informosome: A complex of mRNA and protein that protects mRNA from degradation. (Spirin, 1964)

inheritance: The derivation of characters from its progenitors by an organism.

inhibitor: A substance that blocks the function or growth of an enzyme.

inhibitory epistasis: One gene when dominant is epistatic to the other, but the other when recessive is epistatic to the first. These genes give a 13:3 ratio in F_2 progeny.

initiation codon: The codon, usually but not exclusively 5′-AUG-3′, that indicates the point at which translation of an mRNA should begin.

initiation complex: The complex that comprises mRNA, a small ribosomal subunit, aminoacylated initiator-tRNA, and initiation factors and that forms during the initiation stage of translation. In prokaryotes, the complex contains a 30S ribosomal subunit, transfer RNA for formylated methionine ($tRNA_f^{met}$), messenger RNA, and initiation factors. The initiation complex is stabilized by base pairing between the 3′ terminal sequence of 60S ribosomal RNA (. . . CCUC-CUUA) and a purine-rich nucleotide stretch located in mRNA on the 5′ side of the AUG initiator. A 50S ribosomal subunit is then added to form a 70S initiation complex. In eukaryotes, it is a ternary complex containing initiator tRNA that interacts with a 40S ribosomal subunit to form a 40S preinitiation complex. Subsequently, mRNA binds to the preinitiation complex, and the 80S initiation complex is formed by joining the 60S ribosomal subunit to the 40S intermediary complex in which the $tRNA_f^{met}$ is at the P-site. (Nomura and Ray, 1980)

initiation factors: Protein molecules that play an ancillary role in the initiation stage of translation. In prokaryotes, three IFs have been recognized. The formylmethionine-tRNA binding factor (IF2), the mRNA binding factor (IF3), and the factor involved in recycling IF2 (IF1). In eukaryotes, nine IFs have been observed. eIF3 is the largest and structurally the most complex factor that binds to the small ribosomal subunit. At least five eIFs are involved in peptide chain initiation.

initiation site: The site on the DNA molecule where the synthesis of RNA begins.

initiator proteins: Proteins that recognize the origin of replication on a replicon and take part in primosome construction. (Jacob et al., 1960)

initiator tRNA: A specific form of RNA that carries a free or formylated methionine residue and interacts with the initiation codon to start the synthesis of a protein molecule. The methionine, which is essential for such initiation, is removed from most proteins before or shortly after synthesis is terminated. Also known as *initiating tRNA*.

inosine I: A newly discovered nucleotide that is found in the third position in a codon and can pair with A, U, and C, resulting in wobble base pairing. It is a deamination product of adenosine.

input load: The load of inferior alleles in a gene pool caused by mutation and immigration.

insert: (1) A length of DNA that is linked into another length of DNA or a cloning vehicle using gene manipulation techniques; (2) the piece of foreign DNA introduced into a vector molecule.

insertion: A mutation in which one or more new bases are added between preexisting bases on a nucleic acid chain.

insertion elements: Sequences of DNA that specify the location of insertion of different episomes into the genome.

insertion mutagenesis: Change in gene action due to an insertion event that either changes a gene directly or disrupts control mechanisms.

insertion mutation: A mutation caused by the insertion of an extra base or a mutagen between two successive bases in DNA.

insertion region: The centromere, spindle atttachment, or kinetochore; a nonstaining, localized region in each chromosome to which the spindle "fiber" appears to be attached at metaphase. The centromere remains single for some time after the rest of the chromosome has divided and, at anaphase, starts to move toward the pole before the rest of the chromatid.

insertion sequences (IS elements): Any of a class of distinct prokaryotic segments of DNA (generally shorter than 2 kb) that are able to transpose to numerous sites on bacterial plasmids, chromosomes, and bacteriophages. They contain no transposase gene. They can modulate gene expression and cause rearrangements of genomes. *See* TRANSPOSABLE GENETIC ELEMENT.

insertion (cloning) site: A unique restriction site in a vector DNA molecule into which foreign DNA can be inserted. The term is also used to describe the position of integration of a transposon or IS element.

insertion vector: A phage-cloning vector, usually λ, that has unique restriction sites in a nonessential region. λ phages need to be between 75 and 105 percent of the size of wild-type λ DNA to be packed, and insertion vectors are at the lower end of these size limits. These vectors therefore can accommodate about 20 percent of the size of wild-type λ (i.e., up to 10 kb of insert DNA). The site into which foreign DNA is cloned is often within a gene that enables recombinant phage plaques to be distinguished from plaques by phages that contain no insert DNA.

insertional inactivation: The technique where foreign DNA is cloned into a restriction site that lies within the coding sequence of a gene in the vector. The insertion of foreign DNA at such a site interrupts the gene's sequence such that its original function is no longer expressed. This permits the detection of recombinant molecules following transformation.

insertional mutagenesis: Mutation or alteration of a DNA sequence as a result of the insertion of DNA.

insertional translocation: The insertion of a segment from one chromosome into another nonhomologous one.

inside marker: The middle locus of three linked loci.

integrase: A viral enzyme that catalyzes site-specific recombination involving the integration of the genome of a temperate bacteriophage into a bacterial host chromosome. (Zissler, 1967)

integration efficiency: The frequency with which a particular DNA sequence is incorporated into the genome of a recipient bacterium following transformation, or into a recipient animal or plant cell or ovum following its artificial introduction.

integrator gene: A component of eukaryotic gene regulation that responds to the signals provided by sensor genes. These genes are transcribed to produce activator RNAs. (Britten and Davidson, 1969)

interaction of genes: The process by which one gene-difference affects the expression of another gene-difference.

interaction theory: A theory of quantitative factor inheritance which assumes that the effect of each factor on the genotype is dependent upon all the other factors present. The visible effect of a certain factor is smaller the greater the number of factors acting in the same direction.

interallelic complementation: The change in the properties of a heteromultimeric protein brought about by the interaction of subunits coded by two different mutant alleles; the mixed protein may be more or less active than a protein consisting of subunits only of one or the other type.

intercalary: A term applied to the region between specified sites of the chromosome. Also known as *interstitial.*

intercalary heterochromatin: Heterochromatin, other than centromeric heterochromatin, dispersed throughout eukaryotic chromosomes.

intercalary segments: Indicated segments of chromosomes between the terminal segment and centromere in each arm.

intercalating agents: Certain drugs or dyes are able to invade the space between adjacent base pairs of a double-stranded DNA molecule, often causing mutations. Molecules with this property are called intercalating agents. The binding of such dyes reduces the buoyant density of the DNA. The DNA duplex increases in length, and if the DNA is supercoiled, increasing concentrations of the dye first unwind the supercoils and then wind the molecule up again in the opposite position. A mutagen that inserts itself between two successive nucleotides and causes a frameshift mutation is known as an *intercalating mutagen*.

intercalation: (1) Insertion of a molecule (e.g., acridine dye) between base pairs of accurate base pairing and alignment in the DNA double helix; (2) growth of tissue to replace missing parts of appendages following incomplete grafting (e.g., if a tibia is grafted onto the proximal part of a femur); (3) growth of meristematic tissue in plants between two other differentiated tissues.

interchange: An exchange of nonhomologous terminal segments of chromosomes. (Belling, 1925)

interchange complex: The association of four or more chromosomes in a translocation heterozygote at diakinesis or metaphase I.

interchromosomal recombination: Recombination resulting from independent assortment.

interchromosome balance: The condition in which whole chromosomes are adjusted in proportions that allow for satisfactory development of the organism.

intercistronic region: The distance between the termination codon of one gene and the initiation codon of the next gene.

intercross: A cross between two heterozygotes for the same alleles.

interdeme selection: Selection at the level of a local population.

interference, chromatid: The nonrandom involvement of the chromatids in a double crossing-over. (Mather, 1933)

interference microscope: A light microscope that measures the refractive index of microscopic objects, including parts of living cells. It is similar to a phase-contrast microscope but also provides a guide to dry mass or thickness.

intergenic changes: Mutational changes involving more than one locus (e.g., inversions, deletions, translocations, duplications).

intergenic region: The region of DNA between gene-coding sequences.

intergenic (extragenic) suppression: A mutation at a second locus that apparently restores the wild-type phenotype to a mutant at a first locus. Intergenic suppressors are of three types: nonsense, missense, and frameshift.

intergroup selection: An evolutionary process due to differences in growth rate or survival of different competing or noncompeting groups.

interkinesis: Interphase or the resting stage between two divisions of a cells but not applied to resting stages in general.

intermediate component: Components of a reassociation reaction are those reacting between the fast (satellite DNA) and slow (nonrepetitive DNA) components; they contain moderately repetitive DNA. (Gregoire, 1905)

intermediate (blending) inheritance: In the inheritance of quantitative character, the F_1 generation is often intermediate between the parents and the F_2. This intermediate character may apparently prevail. This phenomenon is due to polymeric genes. *See* POLYMERISM.

intermitosis: The resting stage that often occurs between two miotic divisions—interphase.

internal control region (ICR): In all eukaryotic genes that are transcribed by RNA polymerase III, a small DNA sequence directs upstream initiation by the enzyme. ICRs occur within the DNA encoding the mature RNA. Eukaryotic tRNA gene transcription is controlled by two ICRs, the D-control region (which in the tRNA codes for the D-stem and -loop), and T-control region (which in the tRNA codes for the T-loop). (Skonju, Bogenhagen, and Brown, 1980)

internal genetic balance: The harmonious epistatic interactions of genes at different loci.

internal guide sequence (IGS): In group I introns, a sequence pairing with exon sequences adjacent to both the 5′ and 3′ splice sites. The IGS usually begins with twenty nucleotides of the 5′ splice site. IGS 5′ exon pairing is important in both steps of RNA splicing. (Davies, Fuhrer-Krusi, and Kucherlapati, 1982)

internal promoter: Promoter sequence that occurs within the coding sequence of a gene. The 5S genes of *Xenopus* are known to have an internal promoter in addition to the conventionally placed promoter upstream of the 5′ end of the gene.

interphase: The part of the eukaryotic cell cycle during which chromosomes are decompacted as chromatin and enclosed in the nuclear envelope. Metabolically, a very active stage, and chromosomes replicate this phase. It is subdivided into G_1, S, and G_2 subphases. (Lundegardh, 1912)

interphase mapping: A technique that involves the *in situ* hybridization of labeled DNA probes to the interphase chromatin. It provides the opportunity for mapping at a much higher resolution. The limit of resolution in interphase mapping ranges only 25 kb to 100 kb. (Lawrence, 1990)

interspecific selection: Selection that operates to improve the competitive power of one species in relation to other species, as distinct from *intraspecific selection.*

interspersion pattern of DNA: The pattern of distribution of coding and noncoding sequences within a genome. In many genomes, especially of higher eukaryotes, there is a tendency for coding sequences of structural genes to be interspersed with long lengths of DNA that are not transcribed. Such DNA is partly regulatory but much may be redundant.

inter-SSR-PCR: A technique closely related to MP-PCR (microsatellite-primed PCR) that uses 5′ or 3′ anchored microsatellites of the dinucleotide repeat type as PCR primers (e.g., $GG[CA]_7$). This technique is also called *AMP-PCR (anchored microsatellite primed PCR),* during which complex fingerprint-like patterns are obtained. The banding pattern complexity and informativeness of AMP-PCR is considerably higher as compared to MP-PCR (Microsatellite Primed PCR).

interstitial chiasma: A chiasma that has a length of chromatid on either side of it.

interstitial region: The chromosomal region between the centromere and the site of a rearrangement.

interstitial segment: Segments of chromosomes in interchange hybrids, which are nonhomologous with end segments.

intervening sequences: In a discontinuous gene, one of the segments that does not contain biological information. A sequence present in a gene but not present in its mRNA. The more commonly used term is *introns*. (Jeffreys and Flavell, 1977)

intra-allelic competition: The restoration of activity to an enzyme made of subunits in a heterozygote of two mutants that, when homozygous, do not have any activity. This is caused by interaction of the subunits in the protein.

intra-allelic complementation: Synergistic effect observable in some proteins that are mixed polymers with subunits coded by more than one allele or more than one gene. Thus, mixed tetramers of haemoglobin and lactate dehydrogenase are more efficient in some situations than are the homopolymers.

intracellar promiscuity: Intercompartmental gene transfer of DNA. *See* PROMISCUOUS DNA. (Ellis, 1982)

intrachromosomal recombination: Recombination resulting from crossing-over between two gene pairs.

intrachromosome balance: A balance produced by the adjustment of the genes within one chromosome.

intracistronic complementation: A type of complementation between mutant variants of certain genes encoding catalytic enzymes, repressor type proteins, or membrane proteins. Complementation is the result of the combination of mutant polypeptides in which each mutant chain provides or corrects the conformation and the site(s) of activity of the protein. (Jackson and Yanofsky, 1974)

intraclass correlation: A measure of phenotypic resemblance between relatives that is determined by the ratio of covariance of full sib (FS) or half sib (HS) to the total phenotypic variance (V_p), i.e., $t(FS)/V_p$ and $t(HS) = Cov(HS)/V_p$. The higher the size of t, the greater the resemblance between relatives.

intracodon recombination: Genetic recombination between adjacent nucleotides within a codon.

intragenic changes: Mutational alterations in the nature of the individual gene, as opposed to intergenic changes.

intragenic mutations: Mutations that occur within the same gene.

intragenic suppression: When the effect of a mutation in one gene is completely or partially suppressed by another mutation in the same gene.

introgression: The incorporation of genes from one species into the gene pool of another via an interspecific hybrid. (Anderson and Hubricht, 1938)

introgressive hybridization: The crossbreeding of individuals from two different species that results in introgression.

intron: Any of those sequences (intervening sequences) in split genes that interrupt the coding sequence (exons) and are transcribed as a part of a precursor RNA. Introns are excised and exons are ligated by RNA splicing. Nearly all mRNA introns are bound at the 5′ end by a conserved sequence of nine nucleotides called a 5′ splice site and begin with the dinucleotide GU (5′-exon-AAG/GUGAGU-intron-3′). At the 3′ end, they are bound by a 3′ splice site that is a pyrimidine-rich region of about eleven nucleotides followed by N-Y-A-G (N is any nucleotide, Y is pyrimidine). Mitochondrial introns do not obey the GU . . . AG rule. These introns have been classified into two groups. Group I introns have a sixteen-nucleotide-long consensus sequence that is far away from the splice site in the primary sequence. Group II introns include consensus sequences at the 5′ and 3′ ends. They have a characteristic 14 bp potential hairpin. Mitochondrial introns contain open reading frames. (Gilbert, 1978)

intron-exon junction: A DNA sequence at the boundaries of introns and exons of mosaic genes. These regions constitute the basis for specifying the site of RNA splicing.

intronic gene: A gene located inside an intron and surrounded by noncoding regions.

invariance: The reciprocal of the variance.

inversion mutation: Alteration of the sequence of a DNA molecule through removal of a segment, followed by its reinsertion in reverse orientation. (Sturtevant, 1926)

inversion, overlapping: A compound inversion caused by a second inversion that includes part of a previously inverted segment.

inversion polymorphism: The presence of two or more chromosome sequences, differing by inversions, in the homologous chromosomes of a population.

inverted repeat sequence: This sequence is comprised of two copies of the short sequence of DNA repeated in opposite orientation on the same molecule. Inverted repeat sequences are present in the chromosomal DNA of many prokaryotes and eukaryotes and involve the promoter regions, the recognition sites of restriction endonucleases, and the repressor binding sites. Adjacent inverted repeats may form double hairpin or double stem and loop structures (palindromes). Possible functions of inverted repeats are in the initiation of DNA replication, transcription termination, and transposition. (Davidson et al., 1973)

inverted terminal repeats: The short, related or identical sequences present in reverse orientation at the ends of some transposons.

iojap: A locus in corn that produces variegation.

ion exchange chromatography: A laboratory of full-scale industrial process in which ion exchanges are used to fractionate mixtures of charged molecules that may be inorganic, low molecular weight, organic compounds or macromolecules.

ion exchange column: A column packed with material suitable for ion exchange chromatography. Such materials are resins, cellulose, or other supporting compounds to which charged groups have been attached. For example, diethlaminoethyl-cellulose (DEAE-cellulose) is positively charged at pH 7.0 and so, acts as an anion exchange material; carboxymethyl-cellulose (CM-cellulose) is negatively charged and used for cation exchange chromatography. The mixture of compounds (e.g., proteins) having different isoelectric points, and thus different charges at a given pH value, is passed down the column material and eluted sequentially by a gradient of buffer of increasing ionic strength, or by a gradient of changing pH value.

irradiation breeding: *See* MUTATION BREEDING.

IS element: *See* INSERTION SEQUENCES. (Hirsch, Starlinger, and Brachet, 1972)

isoacceptor tRNAs: Two or more tRNA molecules that are specific for the same amino acid. (Doctor, Apagar, and Holley, 1961)

isoalleles: Alleles with similar phenotypic effect. (Stern and Schaeffer, 1943)

isochromatid break: A break, usually induced by X rays or other ionizing radiation, at the same locus in two already divided chromatids. (Sax, 1941)

isochromosomal lines: Contrasted with clones, homozygous lines, or identical twins, which are identical at all loci, the isogenic lines differ from each other at one locus only. For instance, A-line (male sterile) and corresponding B-line (maintainer) are isogenic differing only for fertility locus.

isochromosome: A chromosome with two genetically and morphologically identical arms. The arms are mirror images of each other. (Darlington, 1940)

isoelectric focusing: A method of separating proteins according to their net electric charge. It is usually performed in gels, as in conventional electrophoresis, but a range of special Zwitter ionic-buffering agents are used so that, under the influence of an electric field, a pH gradient is set up in the gel. Proteins move in the electric field until they reach a pH that is the same as their isoelectric points (PI), at which point they are uncharged and they remain at that point in the gel. This method can be used preparatively for purifying proteins in a mixture of very similar isoelectric points or for very high-resolution analytical work.

isogenic: Having the same genotype. (Johannsen, 1926)

isogenic line: A group in which all members have identical, hereditary makeup (e.g., clone, homozygous inbred line, identical twins).

isogenic stocks: Strains of organisms that are genetically uniform but not necessarily homozygous.

isolating chromosomal mechanism: A difference in chromosome number or morphology that maintains genetic isolation or lack of crossing.

isolating mechanisms: Any structural, physiological, behavioral, or other feature of an individual, or any geographical or geological barrier that prevents individuals of one population from successfully interbreeding with those of other populations.

isolines: Lines that are genetically similar except for one gene.

isomorph: *See* HYPOMORPH.

isophene: Lines on geographic maps that connect points of equality in the phenotypic expression of a clinally varying character. (Mayr, 1963)

isoschizomers: Restriction endonucleases that recognize the same target sequence and cleave it the same way.

isotelo compensating trisomic: A compensating trisomic in which the missing chromosome is compensated for by one isochromosome and one telocentric chromosome. (Kimber and Sears, 1968)

isotertiary compensating trisomic: A compensating trisomic in which the missing chromosome is compensated for by one isochromosome and one tertiary chromosome. (Kimber and Sears, 1968)

isozyme markers: Markers that detect allelic variants of specific enzymes.

isozymes: Different forms of the same enzyme that catalyze particular biochemical reactions during metabolism. (Markert and Møller, 1959)

I-value: The total concentration of DNA terminal in a ligation reaction, which increases with increasing DNA concentration. I-values and J-values influence the formation of products in a ligation reaction. By choosing different I- and J-values (i.e., various concentrations of vector and insert DNAs) one can influence the structures of the recombinant molecules formed. These may be linear oligomers, circularized vectors, or vectors plus insert.

Jacob and Monod model: The operon model of gene regulation in the *lac* region of *E. coli*. This model explains the mechanism of gene expression in prokaryotes. *See* GENE REGULATION MECHANISM IN PROKARYOTES. (Jacob and Monod, 1961b)

joint DNA molecule: Any DNA molecule containing material contributions from two different DNA duplexes, as a result of breakage and reunion recombination *(in vivo)* or due to recombinant DNA technology *(in vitro).*

juggernaut polymerase: A transcriptase (DNA-dependent RNA polymerase) that ignores all signals for termination of transcription due to the rho factor. This polymerase is assumed to originate by attachment of other proteins conferring antitermination ability. (Adhya, Gottesman, and de Crombrugghe, 1974)

jumping gene: The gene that keeps on changing its position in a chromosome and also between the chromosomes in a genome. *See* TRANSPOSABLE GENETIC ELEMENT.

junctional protein: Proteins specifically associated with cell junctions.

junctional sequence: Any of the two terminal regions of the intron in RNA precursors.

junk DNA: *See* SELFISH DNA.

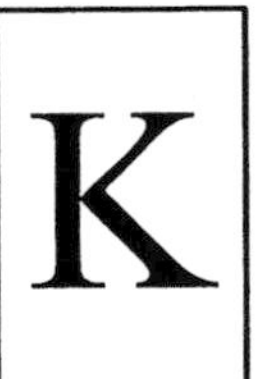

kanamycin: Antibiotic that operates through its ability to bind to bacterial ribosomes and causes misreading of the genetic message.

karyokinesis: The division of the nucleus during cell division. (Schleicher, 1878)

karyological races: Races of species with all members having the same chromosome number, which differs from that typical for that species.

karyolymph: A clear fluid or nuclear sap within the nuclear membrane.

karyoplast: A cell nucleus, obtained from the cell by enucleation, surrounded by a narrow rim of cytoplasm and a plasma membrane.

karyosphere: The part of the nucleus into which the chromosomes have contracted.

karyotype: A set of characteristics in respect to chromosome number, the shape and relative size of the chromosomes, number of secondary constrictions, absolute size of the chromosomes, and distribution and size of heterochromatic and euchromatic regions of a particular chromosome complement of an individual or of a related group of individuals. (Levitsky, 1924)

killer yeast: Yeast cells containing a double-stranded RNA plasmid that encodes genes responsible for production of a killer substance that is the lethal to other cells. The substance is closely similar in its action and synthesis to a bacterial colicin.

kilobase (kb): One thousand nucleotides or nucleotide pairs in sequence. May be used to pertain to either DNA or RNA.

kin selection: The mode of natural selection that acts on an individual's inclusive fitness. (Hamilton, 1964)

kinase: An enzyme that will add a phosphate group to a protein or nucleic acid; the process known as *phosphorylation*.

kinetic complexity: The complexity of a DNA component measured by the kinetics of DNA reassociation.

kinetic constriction: Centromere; spindle attachment; insertion region; kinetochore; a nonstaining, localized region in each chromosome that remains single for some time after the rest of the chromosome has divided and that, at metaphase, appears to be attached to the spindle fiber.

kinetics: A dynamic process involving motion.

kinetin: A synthetic cytokinin. Cytokinins are a group of plant growth substances, natural or synthetic, that stimulate cell division.

kinetochore: A proteinaceous disc bound to centromeric DNA to which microtubules of the spindle apparatus attach. (Schrader, 1936)

kinetoplast: These are modified mitochondria located near to the base of flagella. A kinetoplast may be larger than a nucleus. (Alexeiff, 1917; Lwoff, 1949)

kinetoplast DNA: DNA present in kinetoplasts. This DNA exists in the form of a large number of interlocked "minicircles." (Simpson, 1972)

kink: An abrupt and significant bend between two helical segments of DNA achieved by unstacking one base pair and twisting (kinking) the polynucleotide backbone. (Crick and Klug, 1975)

"kissing" in RNA pairing: The first step in RNA pairing, which involves initial loop-loop interaction.

Klenow fragment: The larger part of the bacterial DNA polymerase I (76KD) that remains after treatment with subtilisin; retains some but not all exonuclease and polymerase activity.

knob: A large, heavily staining, bead-like structure at or near the ends of pachytene chromosomes of certain organisms (e.g., maize). Knobs may be used as cytological markers (such as chromomeres) for the identification of specific chromosomes. (Longley, 1937)

Konzak's scanning hypothesis: Pertains to translation initiation in eukaryotes. Accordingly, a 40S ribosomal subunit associated with met-tRNA moves down the mRNA until it encounters the first AUG, which in 90 percent of the cases occurs in the form of consensus sequence PuNNAUGG, known as *Konzak's consensus sequence.*

Kornberg enzyme: DNA polymerase discovered in cells of *E. coli.* It catalyses the addition of mononucleotides to the 3′ end of a growing DNA chain provided *in vitro* with a DNA template, the four deoxyribonucleoside triphosphates, and Mg^{2+}; same as *DNA polymerase I.* (Kornberg, Lehman, and Schims, 1956)

Kosambi function: The recombination values cannot be converted into map distances without further accounting for double crossovers. A formula has been given by Kosambi which assumes that the coincidence—ratio of observed frequency and expected frequency of double crossovers—without interference, itself depends linearly on the recombination rate. (Kosambi, 1944)

Kozak sequence: In the 5′ untranslated mRNA region (UTR), a base sequence (CCRCCATGG; R = purine) that appears to be required for the most efficient recognition of the correct initiator codon by the ribosomes in eukaryotic cells.

Krebs cycle: An enzyme system that converts pyruvic acid to carbon dioxide in the presence of oxygen, with a concomitant release of energy that is captured in the form of ATP molecules. Also referred to as *citric aid cycle* or *tricaboxylic acid (TCA) cycle.*

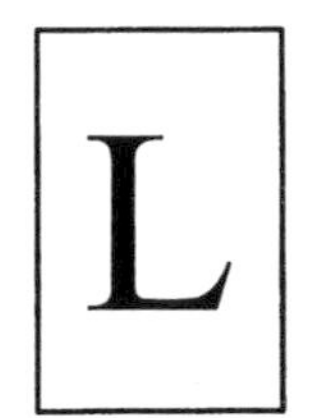

labile gene: Genes that are constantly mutating.

***lac* operon:** The cluster of three structural genes that code for enzymes involved in utilization of lactose by *E. coli.*

lactose repressor: The regulatory protein that controls transcription of the *lac* operon in response to the levels of lactose in the environment.

ladder system approach: A systematic approach in plant breeding commencing from developing genetic base, exploitation of variability, removal of bottleneck genes, boosting the productivity mechanism, quality improvement, adaptive, and finally, leading to the release of an improved variety. This is a ladder system approach in plant breeding.

lag phase: The period of slow, orderly growth when a medium is first inoculated with a culture.

lagging chain: The strand of the double helix that, during DNA replication is synthesized in a discontinuous fashion in the form of Okazaki fragments. *See* OKAZAKI FRAGMENT.

Lamarckism. The theory of evolution proposed by Lamarck that espouses the inheritance of acquired characteristics. *See* THEORY OF INHERITANCE OF ACQUIRED TRAITS.

land races: Prescientific local cultivars which are the products of natural selection over time under domestication.

lariat RNA: An RNA intermediate generated during splicing of mRNA precursors. The lariats have a circular component with an extending tail and are formed by a branch where an adenosine residue is linked to the 5′ end of an intron and to the remainder of the intron. (Padgett et al., 1984)

Latin square design: An experimental design that simultaneously controls fertility variation in two directions.

lattice design: An experimental design in which the number of treatments forms a square.

Law of Ancestral Inheritance: The idea that an individual contains, in equal amounts, genetic material from each of the four grandparents, each of the eight blends of the previous generation, and that each ancestor contributes a specific proportion of an individual's genome.

LD (lethal dose): (1) The number of pathogenic microorganisms required to cause death in a given species of animal or plant. (2) Radiation required to kill within a specified time all the individuals in a large group of organisms.

LD_{50} (lethal dose-50): The dose of mutagen (physical or chemical) causing 50 percent mortality of the target material.

leader DNA: The untranslated segment that lies upstream of the initiation codon on an mRNA molecule.

leader peptide: A short peptide synthesized *in vitro* by translation of the leader sequence of some bacterial mRNAs. These peptides are not formed *in vivo*.

leader peptide gene: A small gene within the attenuator control region of repressible amino acid operons. Translation of the gene tests the concentration of amino acids in the cell.

leader transcript: The untranslated segment of mRNA that lies upstream of the initiation codon on an mRNA molecule.

leading strand: The strand of the double helix that during DNA replication is copied in a continuous fashion.

leaky mutant: A mutant gene that produces a protein with partially normal activity.

leaky mutations: A mutation that results in only partial loss of a characteristic.

left splicing junction: The boundary between the right end of an exon and the left end of an intron.

left-handed DNA: *See* DNA HELIX H.

leptotene: A substage of prophase I of meiosis I during which the chromosomes appear thread-like. (Winiwarter, 1900)

lesion: A damaged area in a gene (a mutant site), a chromosome, or a protein.

lethal equivalent alleles: Alleles whose summed effect is that of lethality—for example, four alleles, each of which would be lethal 25 percent of the time (or to 25 percent of their bearers), are equivalent to one lethal allele.

lethal gene: A gene whose phenotype effect is sufficiently drastic to kill the bearer. Death from different lethal genes may occur at any time from fertilization of the egg to advanced age. Lethal genes may be dominant, incompletely dominant, or recessive.

lethal mutation: A mutation that results in a cell or organism that is unable to survive.

leucine zipper: A structural motif found in several DNA-binding proteins. The leucine repeat regions in the protein are capable of forming regular α-helix, with the leucine side chains projecting out from the helix at regular intervals. Leucine side chains of one molecule may interdigitate with those from a second protein, thus forming a zipper that correctly holds the molecules together. Such dimers may be the entities that interact with DNA and regulate gene expression. (Landschulz, Johnson, and McKnight, 1988)

life: (1) A condition that distinguishes plants and animals from inorganic objects and dead organisms. The distinguishing manifestations of life are: growth, metabolism, reproduction, and power of adaptation to environment through changes originating internally. (2) Any entity that is capable of making a reasonably accurate reproduction of itself and duplicate being able to produce the same task and subjected to low rate of alteration. These changes are heritable.

life cycle: The entire series of developmental stages undergone by an individual from zygote to maturity and death.

ligase: *See* DNA LIGASE.

ligation: Formation of a phosphodiester bond to link two adjacent bases separated by a nick in one strand of a double helix of DNA. In a cloning experiment a restriction fragment is often ligated to a linearized vector molecule using T4 DNA ligase. (The term can also be applied to blunt-end ligation and to joining of RNA.)

limited backcrossing: Instead of the complete six cycles of backcrossing (up to BC6), only two or three cycles of backcrossing are coupled with rigorous selection to gain the advantages of transgressive segregation.

"limited" chromosomes: Heterochromatic chromosomes restricted to the germ line in some species.

line: A group of individuals from a common ancestry. A more narrowly defined group than a strain or variety.

line breeding: A system of breeding in which a number of genotypes, which have been progeny tested in respect to some character or group of characters, are composited to form a variety.

linear DNA: A term used to describe the physical state of a nucleic acid molecule. The two strands at the end of a linear molecule of double-stranded DNA can be free, as in a restriction fragment, or bound to a specific protein, as in some eukaryotic extrachromosomes. Eukaryotic chromosomes are linear and their ends, the telomeres, have special sequences that promote their replication.

linkage: The physical association between two genes that results from them being present on the same chromosome. Tendency of the parental combination of genes to stay together in F_2. Linkage can be broken by crossing-over during meiosis. (Morgan, 1910)

linkage analysis: (1) Determination of the relative positions of genes on chromosomes on the basis of their linkage patterns; (2) the use of polymorphic variation to estimate genetic distance.

linkage detection: This deals with the partitioning of total χ^2 into different components to find out presence of linkage.

linkage disequilibrium: The condition among alleles at different loci such that allelic combination in a gamete does not occur as the product of the frequencies of each allele at its own locus. (Kimura, 1956)

linkage equilibrium: The condition among alleles at different loci such that any allelic combination in a gamete occurs as the product of the frequencies of each allele at its own locus.

linkage group: A group of genes that display linkage. With eukaryotes, a single linkage group usually corresponds to a single chromosome. (Morgan, 1911)

linkage map: Graphical representation of the arrangement of genes or genetic markers in a linkage group in such a way that the distance between any two of them reflects the percent of crossing-over between them.

linkage number: The number of times one strand of a helix coils about the other. (Crick, 1976)

linkage value: A recombination percent expressing the proportion of crossovers to noncrossovers (the parental type). The linkage value may vary from 0 to 50 percent, which is the value for independent assortment.

linked genes: Genes that are located on the same chromosome and that, therefore, tend to segregate together.

linker DNA: The DNA that links nucleosomes together and that makes up the "string" in the beads-on-a-string model for chromatin structure. *See* BEADS-ON-A-STRING.

linker insertion mutagenesis: An *in vitro* mutagenesis technique used to generate a large library of mutants with specific alternations that are then scattered throughout a cloned DNA to screen these mutants for a desired phenotype. (Heffron, So, and McCarthy, 1978)

linker oligonucleotide: A synthetic oligodeoxyribonucleotide that contains a restriction site. Linkers may be blunt-end ligated onto the ends of DNA fragments to create restriction sites that can be used in the subsequent cloning of the fragment into a vector molecule. It may be necessary to protect the fragment to be cloned from the action of the restriction endonucleases by treating them with DNA methylase.

linker scanner mutations: Mutations that are introduced by recombining two DNA molecules *in vitro* at a reaction fragment added to the end of each; the result is to insert the linker sequence at the site of recombination.

linking number: The number of times the two strands of a closed DNA duplex cross over each other. (Fuller, 1971)

linking number paradox: The discrepancy between the existence of −2 supercoils in the path of DNA on the nucleosome, compared with the measurement of −1 supercoil released when histones are removed.

localization: The genotypic property of restriction of crossing-over and chiasma formation to certain corresponding parts of all the chromosomes.

locus: The specific place on a chromosome where a gene is located. (Morgan et al., 1915)

LOD score: Log_{10} ratio of two probabilities for recombination ratio, i.e., observed recombination and free recombination. The higher the LOD value, the stronger the linkage. An LOD score of 3 means that the probability of a linkage value is 10^3 compared to the probability of free recombination.

long interspersed nucleotide element (LINE): Any of the long, (6 to 7 kb) repetitious (about 10^4 times), interspersed DNA sequence elements in nuclear genomes, the majority of which are retroposons. LINEs contain one or more open reading frames. (Singer, 1982)

long period interspersion: A pattern in the genome in which long stretches of moderately repetitive and nonrepetitive DNA alternate.

long terminal repeat (LTR): A sequence directly repeated at both ends of retroviral DNA, bacterial and eukaryotic transposons, and in bacteriophage Mu. LTRs carry promoter sequences capable of initiating transcription.

long-range restriction mapping: The construction of long-range maps of genomic DNA by use of restriction endonucleases that cut the DNA infrequently.

long-term gene regulation: Gene regulation recognized in eukaryotes that operates during determination, differentiation, or more generally, development.

loop: A single-stranded region at the end of a hairpin in RNA or single-stranded DNA; corresponds to the sequence between inverted repeats in duplex DNA.

lost gene: A situation in which all the genes at a certain locus are the same, and the other allele has been permanently eliminated from the population.

low pressure liquid chromatography: An automated system for carrying out liquid chromatography.

Ludwig theorem: The theory that new genotypes can be added to a population if they can utilize new components of the environment (occupy a new subniche), even if they are inferior in the ancestral niche.

Luria and Delbrük experiment: An experiment in genetics designed to show whether a specific mutation could be environmentally induced or a random occurrence. It involved the use of a fluctuation test, in which numerous cultures of bacteria were simultaneously screened for resistance to lysis by bacteriophage. If mutation to phage resistance were random, then considerable fluctuation in the numbers of mutant bacteria would be expected in the parallel cultures, depending on precisely when the mutation had occurred (if it occurred early in the life of the culture, many progeny would have accumulated, if late, no progeny would have accumulated). Since marked fluctuations in the numbers of surviving resistant colonies were obtained, it was concluded that mutation was occurring at a constant, random rate and was not being induced by exposure to the phage or other environmental factors. (Luria and Delbrük, 1943)

luxury genes: The genes coding for specialized functions; their products are synthesized (usually) in large amounts in particular cell types.

lysate: The contents released from a lysed cell.

lysis: The bursting of a cell by the destruction of the cell membrane following infection by a virus.

lysogenic bacteria: A strain of bacteria-harboring prophage that causes the lysis of a strain sensitive to that virus. (Lwoff, 1953)

lysogeny: A state in which the genetic material of a virus and its bacterial host are integrated.

lytic cycle: The multiplication cycle of viruses in their vegetative stage that leads to the lysis of the host cell.

lytic cycle control: A mechanism whereby the pathway of transcription of phage DNA is controlled in such a way that either the lytic pathway or the lysogenic pathway is followed. If a protein (cl repressor) is formed lysogeny results, if cro protein is formed progeny develop.

M phase: The period of the cell cycle when mitosis or meiosis occurs.

M I, M II: Symbols for the metaphase of the first and second meiotic division, respectively.

M_1 and M_2, etc: Symbols used to designate the first generation, second generation, etc., following exposure to mutagenic agents (ionizing radiations, chemical mutagens, etc.). *See* R_1, R_2, R_3, ETC.

M13: Widely used cloning vector; single-stranded bacteriophage with a closed circular DNA genome. Since the product of replication of this phage, even when it carries an insert of some other gene, is a single-stranded DNA molecule, these vectors are readily used as templates, for sequence analysis.

maintenance methylation: A process where a half-methylated site in DNA is fully-methylated through the use of *maintenance methylase* enzyme, which transfers a methyl group from S-adenosyl methionine (SAM) to cytosine in eukaryotes and adenine in prokaryotes.

major gene: A gene that may cause sufficiently large variation in the trait being studied as to be easily detected. (Mather, 1941)

major groove: The larger of the two grooves that spiral around the surface of the double helix model for DNA.

male sterility, cytoplasmic: Male sterility resulting from specific cytoplasmic-genic interactions. Sterile cytoplasm denoted by symbols S or cms, in contrast to normal cytoplasm N.

male sterility, genetic: Male sterility resulting from the action of specific genes.

map: Diagrammatic representation of the relative positions of different genes in the chromosomes. There can be genetic (linkage), cytological, or molecular, marker maps.

map distance: Distance is measured as cM (centimorgans), with 1 cM = 1 percent recombination (sometimes subject to adjustments). *See* CENTIMORGAN and MAP UNIT. (Bridges, 1932)

map unit (MU): A unit used to describe the distance between two genes on a chromosome in terms of percent recombination, now superseded by centimorgan. Map units are calculated after taking into account double crossovers. One MU is equivalent to 10^6 bp of DNA. *See* CENTIMORGAN.

map-based cloning: The isolation of important genes by cloning the genes on the basis of molecular maps.

mapmaker: A computer package developed for detection and estimation of linkages. This package is based on the maximum likelihood method and is mostly used for construction of RFLP maps. (Lander et al., 1987)

mapping: The study of the position of genes on chromosomes.

mapping function: The mathematical relationship between measured map distance and actual recombination frequency. (Haldane, 1919)

marker: A locus or allele whose phenotype provides information about a chromosome or chromosomal segment during genetic analysis. *See* CENTROMERE MARKERS; INSIDE MARKER; OUTSIDE MARKERS.

marker DNA: A fragment of known size used to calibrate an electrophoretic gel.

marker gene: A gene that codes for an easily detectable protein. Often used to check the effectiveness of gene transfer and expression.

marker retention: This is a technique used in yeast (a unicellular fungal plant) to test the degree of linkage between two mitochondrial mutations.

marker stock: A stock having some genetically characterized markers.

marker trait: A genetically well-characterized trait in a stock that is useful for genetic analysis.

marker-assisted selection (MAS): The technique that uses molecular markers for selection of the progeny of a cross.

masked mRNA: An mRNA complexed with protein so that it is not translated or enzymatically degraded.

master chromosome: A plant mitochondrial genome that exists in the form of circular DNA molecules of different sizes. It is known as *multipartite genome organization*.

master gene: A gene that controls other genes, particularly when the genes controlled are all characteristically expressed in a particular type of differentiated cell. *See* MASTER-SLAVE HYPOTHESIS.

master-slave hypothesis: This hypothesis explains C-value paradox. A cell has several copies of a gene, one of the copies being the master gene and the other copies being slave genes. (Callan and Lloyd, 1960)

maternal chromosomes: In bisexual organisms, the chromosomes derived from the mother. *See* PATERNAL CHROMOSOMES for comparison.

maternal effect: A nuclear gene product in the cytoplasm of the ova determines the phenotype of the organism. Direction of coiling in snails provides an example of maternal effect.

maternal effect genes: Those genes that affect egg polarity and spacial coordinates of the egg and future embryo.

maternal influence: The influence of the genotype of a heterozygous mother on the eggs and larvae. The dominant color of the mother persists for a short time in eggs and larvae, even with recessive genotype.

maternal inheritance: Phenotypic differences due to factors such as chloroplasts and mitochondria transmitted by the female gamete.

mating type gene: The sequence that controls the choice between two mating types known as α and a; best known in the yeast *Saccharomyces.* Haploid yeast cells occasionally switch their mating type. At a genetic level, such a switch has been shown to involve excision of a particular mating-type gene sequence from the genome, with its replacement by a gene of the alternative mating type.

matriclinous inheritance: Inheritance in which all offspring have the nucleus-determined phenotype of the mother.

matrix: A mass of achromatic material in which chromonemeta are embedded.

maturase: Any of the protein encoded by self-splicing introns that are, together with nuclear-encoded proteins, needed for splicing. All the maturases function only in splicing the intron in which they are encoded (or closely related introns). (Lazowska, Jacq, and Slonimski, 1980; Cech, 1985)

Maxam and Gilbert method: *See* CHEMICAL SEQUENCING.

maximum equational segregation: If the crossing-over always takes place between the gene and the centromere in a multivalent (tri- or tetrasomic), the separation of sister chromatid segments carrying the locus will always take place during the first meiotic division.

maximum likelihood method: This method of estimation of linkage depends on the maximizing of the log likelihood, and hence, of the likelihood function. This method always leads to an efficient statistic and also to a sufficient statistic, if one exists.

mechanical isolation: A mechanism for disallowing interbreeding between members of different populations as a result of structural incompatibility of the male and female secondary reproductive organs.

megabase (mb): One thousand kilobase.

megaevolutionary shift: Refers to new, general adaptations during transspecific evolution.

megaspore: The larger of the two types of haploid spores that are produced by heterosporous vascular plants.

megaspore mother cell (megasporocyte): A diploid cell (2n) that undergoes meiosis to produce four megaspores.

megasporogenesis: The process of the production of megaspores from a megasporocyte.

meiocyte: A cell in which meiosis occurs.

meiosis: The series of events, involving two cell divisions, by which diploid cells are converted to haploid cells. (Farmer and Moore, 1905)

meiosis I: The first division of meiosis during which chromosome number is reduced to one-half. Thus, this is a reductional division.

meiosis II: The second division of meiosis during which chromosome number remains constant. Thus, this is an equational division.

meiospore: A cell that is one of the products of meiosis in plants.

meiotic drive: The tendency of higher organisms to exhibit a differential survival rate of gametes. Also called *segregation distortion.* It acts as a force to alter gene frequencies in natural populations. (Sandler and Novitski, 1957)

melting curve: A graphical plot of the change in absorbance at 260 nm, as a function of temperature that occurs when a solution of DNA is heated. Heating results in denaturation, due to breaking of the hydrogen bonds between the bases and dissociation of the two strands of the double helix. The absorbance of double-stranded DNA is always lower than that of single-stranded DNA, which is lower than that of an equivalent amount of free nucleotides in solution.

melting of DNA: The denaturation of a double-stranded DNA molecule into non–base-paired polynucleotides, possibly although not exclusively by heating.

melting out temperature: The temperature at which nucleic acid duplexes of DNA-DNA or DNA-RNA melt by breakage of the hydrogen bonds between the complementary molecules.

melting profile: A profile, commonly represented on a graph, showing the percentage of a given sample of DNA that melts in a specific time on a gradually increasing temperature scale. Since some DNA, such as imprecisely base-paired sequences, low-molecular weight material, and molecules rich in A/T bases, will melt out at lower temperatures than other DNA, some useful information can be derived from melting profiles.

melting temperature (T_m): The midpoint of the temperature range over which DNA is denatured.

Mendelian character (trait): A character the inheritance originates as result of a primary response and that probably forms the basis for the secondary response.

Mendelian genetics: A branch of genetics that deals with the inheritance of qualitative or oligogenic traits.

Mendelian inheritance: The mechanism of the inheritance of chromosomal genes. Mendel's law of segregation states that genes occur in pairs and segregate from each other during meiosis. Mendel's law of independent assortment states that members of one pair of genes do not influence the way in which other pairs of genes are distributed. (Castle, 1906)

Mendelian population: A natural interbreeding unit of sexually reproducing plants or animals sharing a common gene pool. (Dobzhansky, 1935)

Mendelian ratio: A ratio of progeny phenotypes reflecting the operation of Mendel's laws.

Mendelian segregation: *See* SEGREGATION.

Mendelism: A field of genetic research that is especially concerned with the analysis of gene effects and segregation ratio. (Punnett, 1905)

Mendel's first law: The two members of a gene pair segregate from each other during meiosis; each gamete has an equal probability of obtaining either member of the gene pair. Now known as *principle of segregation*.

Mendel's laws: The principle that hereditary characters are determined by discrete particles (genes) that segregate at random in gamete formation. *See* ELEMENT; FACTOR; PARTICULATE THEORY; MENDEL'S FIRST LAW; MENDEL'S SECOND LAW.

Mendel's second law: The law of independent assortment; unlinked or distantly linked segregating gene pairs behave independently. Now known as *principle of independent assortment (segregation).*

meristem tip culture: The *in vitro* culturing of plant tissue from the meristem tip region for the purpose of regenerating pathogen-free plants.

meristemoid: A localized group of meristematic cells that arise in the callus and may give rise to roots and/or shoots.

merodiploid: A haploid organism that is diploid for a small region of the chromosome, i.e., a partial diploid. Also known as *merozygote*.

merozygote: *See* MERODIPLOID. (Wollman, Jacob, and Hayes, 1956)

Meselson and Radding model: A modification to the Holliday model of chromosome recombination. A twisting and isomerization of strands in the crossover intermediate is proposed, avoiding the requirements in the original Holliday model for nicking in both strands.

Meselson and Stahl experiment: It provides the first evidence for the semiconservative replication of DNA. The protocol used involved labeling *E. coli* DNA with ^{15}N, then abruptly transferring the bacteria to a medium containing only ^{14}N. The distribution of DNA that had caesium chloride density centrifugation and the appearance of a band partially labeled, and therefore of intermediate density, constituted the evidence for a semiconservative pattern of DNA replication. (Meselson and Stahl, 1958)

messenger RNA (mRNA): A transcript of a protein-coding gene. The mRNA from a given organism may vary in size over a range of 8S to 45S. In eukaryotes, mRNAs are usually functionally monocistronic. A 5′ terminal nucleotide sequence preceding the initiator codon (AUG) is termed leader sequence. (Brenner et al., 1961; Jacob and Monod, 1961b)

messenger RNA caps: The addition of m7G group(s) at the 5′ end of most of the eukaryotic primary transcript. These caps are of three types: (1) **Cap O**, a cap with a single methyl group, found in 100 percent of the cases; (2) **Cap 1**, a methyl group may be present on penultimate base at 2′ O position of the sugar moiety, present in most cases; and (3) **Cap 2**, a methyl group may be present in the third base also at the 2′ O position of sugar, present in 10 to 15 percent of the cases.

messenger RNA decay: A process responsible for differential survival of mRNAs in the cytoplasm. Site-specific endonucleases seem to control this process.

messenger RNA splicing: The "editing" of the primary transcription of a gene from a higher organism to excise the noncoding intervening sequences. This leaves a continuous coding sequence specifying the protein in question.

metabolic mutant: A mutant having a lesion in a gene that codes for an enzyme involved in metabolism. Nonmetabolic mutants must, for example, have alterations in parts of their DNA not expressed as enzyme proteins.

metacentric chromosome: A chromosome with a centrally located centromere.

metagenesis: Alternation of generation; an alternation of a sexual with an asexual form, or cytologically, the alternation of a haploid with a diploid stage.

metalloenzyme: A protein associated with a metal atom or complexes of metal atoms and functioning as an enzyme.

metaphase: A stage of mitosis at which the nuclear membrane disappears, spindle apparatus is completely formed, and the chromosomes are most condensed and are arranged on the equatorial plate of the spindle apparatus. (Strasburger, 1884)

metaphase chromosome: A chromosome at the metaphase stage of cell division, when the structure is at its most organized and features such as the banding pattern can be seen.

metaphase mapping: The *in situ* hybridization of labeled DNA probes to the mitotic metaphase chromosomes. It is important for rapid localization to a small chromosome region. The limit of resolution in metaphase mapping is in the 1 mb range. (Lawrence, 1990)

metaphase I: A stage of meiosis I at which the nuclear membrane disappears, the spindle apparatus is completely formed, and the homologous chromosomes, in the form of bivalents, are arranged on the equatorial plate of the spindle apparatus.

metaphase pairing index: The proportion of metaphase cells of which two particular chromosomes have paired, to the total number of cells examined.

metaphase plate: The plane of the equator of the spindle where chromosomes are positioned during metaphase.

metaphase II: A stage of meiosis II at which the nuclear membranes in the two haploid cells seen at prophase II disappear, the spindle apparatus is completely formed in each haploid cell, and the chromosomes are arranged on the equatorial plates of the spindle apparatus.

metaxenia: The influence of pollen on the maternal tissues of fruit. *See* XENIA EFFECT.

methyl blue: A dye, normally dissolved in ethanol, that is used in histology for staining cell nuclei.

methyl green: The basic dye used in histochemistry to stain DNA. It is often used in conjunction with pyronine.

methyl methanesuphonate: A frequently used, very potent chemical mutagen that acts by adding methyl to guanine and subsequently causes base pairing errors as it binds to adenine.

methylase: A DNA modification enzyme that covalently attaches a methyl (CH^-_3) group to a specific nucleotide base within a DNA molecule.

methylated cap: A methylated guanine nucleotide that is added to the 5′ end of eukaryotic hnRNA and mRNA after transcription but prior to later processing and translation. *See* CAPPING.

methylation: The attachment of a methyl group to some molecule, such as a histone, RNA, or DNA. *See* DNA METHYLATION.

methylene blue: A histological stain distinct from methyl blue but fulfilling a similar function.

methylmercuric hydroxide (CH_3HGOH): A powerful denaturing agent that may be incorporated into agarose gels in order to determine the single-strand molecular weight of RNA or DNA molecules.

methyltransferase: An enzyme that transfers a methyl group from S-adenosyl methionine to a substrate. Encountered in posttranslation modification of proteins and nucleic acids, in the removal of methyl groups added to DNA by alkylating carcinogens, and in bacterial chemotaxis where the methyl-accepting chemotaxis proteins (MCPs) become methylated in the course of adaptation.

metrical variation: *See* CONTINUOUS VARIATION.

mic RNA: Messenger interfering complementary RNA, i.e., RNA molecules able to bind to the transcripts of particular genes and consequently prevent their translation. (Mizuno, Chou, and Inouye, 1983)

Michaelis constant (Km): Concentration of substrate that permits an enzyme reaction rate to be half-maximal, or at which half of the substrate molecules are engaged with enzyme molecules. *See* ENZYME KINETICS.

microcell: A cell fragment, containing one to a few chromosomes, that is formed by the enucleation or disruption of a micronucleated cell. (Ege and Ringertz, 1974)

microchromosomes: Small chromosomes that pair at meiosis only at metaphase and therefore without chiasma formation.

microevolution: Evolutionary events usually viewed over a short period of time, such as changes in gene frequency within a population over a few generations. (Philipschenko, 1927)

microfilaments: Thread-like organelles involved in cell motion.

microinjection: A micromanipulation technique in which part of one cell is injected into another. The technique is used to inject the nuclei of sperm into cells or nuclei or to inject organelle into cells.

micrometer (μm): A commonly employed unit of measurement in microscopy, it equals 1×10^{-6} meter or 1×10^{-3} millimeter. Also known as *micron* (μ) in older usage.

micromutation: Mutations without visible phenotypic changes, generally observed for quantitative characters.

micron: A unit of length convenient for describing cellular dimensions; it is equal to 10^{-3} cm or 10^{5} Å.

micronucleated cell: A cell that has been mitotically arrested and in which small groups of chromosomes function as foci for the reassembly of the nuclear membrane, thus forming micronuclei, the maximum of which would be equal to the total number of chromosomes.

micropropagation: *In vitro* clonal regeneration of plants from isolated meristematic cells or tissues, usually with an accelerated proliferation of shoots during subcultures.

microsatellites: It is a class of hypervariable loci consisting of tandemly repeated, short (less than 10 bp), simple sequences of 100 to 200 bp length distributed in eukaryotic genomes. These sequences serve as molecular markers for selection of agronomically useful traits and for preparation of advanced, high-density, genetic linkage maps. Microsatellites, also called *simple sequence repeats (SSRs), simple tandem repeats (STRs),* or simply *simple sequences (SSs),* consist of head-to-tail tandem arrays of short DNA motifs (usually 1 to 5 bases). They are a common component of eukaryotic genomes but are almost absent from prokaryotes. Microsatellites are (a) highly variable due to a variable number of tandem repeat (VNTR) -type of polymorphism, (b) more or less evenly distributed throughout genome, and (c) probably nonfunctional and therefore, selectively neutral. (Litt and Lutty, 1989)

microsome: Fragment pieces of endoplasmic reticulum associated with ribosomes.

microspore: In plants, a haploid cell (meiospore) that gives rise to a pollen grain (male gametophyte).

microspore mother cell: A diploid cell that goes through two divisions—one a reductional—producing four microspores. It is the same as the pollen mother cell.

microsporogenesis: The process whereby microspores are produced from pollen mother cells. Meiosis produces four microspores that, when mature, are the pollen grains.

microsurgery by radiation: The breaking of a chromosome in two places, and the insertion of a fragment into another chromosome that has likewise been broken.

microtubule: A hollow, tubular, cytoplasmic component (outside diameter about 15 to 30 nm) found in the cytoplasm, especially of motile cells. It forms the spindle of the mitotic apparatus. (Slautterback, 1963)

mid-parent value: The mean of the values of a quantitative phenotype for two specific parents.

mid-prophase: The stage following zygotene when paired chromosomes are somewhat shortened and thickened, and a stage at which chromosomes have been studied extensively. Also called pachytene.

migration coefficient: The constant proportion of individuals migrating from one population to another. Also known as *coefficient of migration.*

mimic gene: Two or more nonallelomorphic genes that produce similar or identical effects.

mini Ti plasmid: A derivative of the wide-type, tumor-inducing plasmid of *Agrobacterium tumefaciens,* in which part of the DNA sequence that is not needed for infection and replication is excised. This leaves a smaller plasmid that may be used as a cloning vehicle for recombinant gene technology in higher plants.

minichromosome: A chromosome of SV40 or polyoma is the nucleosomal form of the viral circular DNA. (Griffith, 1975; Yasuda and Hirota, 1977)

minimal life: Sequencing of the *Mycoplasma genitalium* genome represents the first complete molecular definition of minimal life with 482 genes. The genome is only 580 kb long. (Goffeau, 1995)

minimum descriptor: The minimum number of distinct morpho-physiological features that can effectively discriminate among genotypes for evaluation of germplasm collections.

mini-preparations: DNA prepared from small bacterial cultures derived from individual colonies in a cloning experiment.

minisatellites: These are families of about 15 bp repeats forming 0.5 to 30 kb sequences distributed throughout the eukaryotic genome. These show allelic differences in the number of repeats called variable number of tandem repeats (VNTRs). The occurrence of many highly polymorphic DNA loci simultaneously is responsible for the specific DNA fingerprints of different individuals. (Jeffreys, Wilson, and Thein, 1985)

minor gene: A gene whose effect on a given trait is so small that it is not easily detected. *See* POLYGENES.

minor groove: The smaller of the two grooves that spiral around the surface of the double helix model for DNA.

minus ten (− 10) sequence: *See* PRIBNOW BOX.

minus thirty-five (− 35) sequence: A region of DNA upstream of prokaryotic promoters that is centered about 35 nucleotides from the mRNA initiation site. The − 35 sequence is thought to be involved in the initial recognition between RNA polymerase and the promoter site. The conserved sequence in the − 35 region of most prokaryotic promoters is:

5′	T	T	G	A	C	A	3′
3′	A	A	C	T	G	T	5′
	−36	−35	−34	−33	−32	−31	

minus twenty-five (−25) box: A component of the nucleotide sequence that makes up the prokaryotic promoter.

misalignment mutagenesis: A spontaneous, mutation-induction process in which the bases are correctly paired, but the pairing occurs out of register.

misdivision: Spontaneous, crosswise (instead of lengthwise) division of the centromere on the spindle, especially of univalents at *anaphase I* and daughter univalents at *anaphase II. See* TELOCENTRIC CHROMOSOME. (Darlington, 1939b)

misdivision haploid: Individual product due to the misdivision of the telocentric or isochromosomes in the haploid.

mismatch DNA repair: A form of excision repair initiated at the sites of mismatched bases in DNA.

mispairing: The presence in one nucleotide chain of a DNA molecule of a nucleotide which is not the complement of that at the corresponding position in the other chain.

missense intergenic suppressor: It is a gene coding for a mutant tRNA able to respond to one or more of the missense mutations. (Yanofsky, Helinski, and Maling, 1961)

missense mutation: An alteration in a nucleotide sequence that converts a codon specifying one amino acid into a codon for a second amino acid. (Brenner et al., 1961)

mistranslation: There is some error rate for normal translation. This value is estimated to be one codon out of every 300 read. Under conditions of amino acid limitation and tRNA imbalance, the error rate increases.

mitochondrial complementation: The heterotic effect expressed in terms of greater mitochondrial activity (ATPase, oxidative activity) of the mixture of mitochondria from two different inbreds than that of the inbreds themselves. This may be a laboratory exercise to identify potential inbred parents for hybridization.

mitochondrial (mt) DNA: In most eukaryotic cells mtDNA is double-stranded and circular (as in bacteria) in structure. It replicates independently of nuclear DNA in a semiconservative manner.

Molecular weight equals 9×10^6 to 10×10^6; length equals 5 μm; size equals 14,000; size 15 to 18 kb in mammals and 208 to 2400 kb in land plants. The master circular chromosome in mitochondria of plants may be resolved into two or more subgenomic circles by intramolecular recombination. The master chromosome exists in equilibrium with subgenomic circles. Mitochondrial genomes code for 3 rRNAs (26S, 18S, 5S) in plants and 2 in humans (16S, 12S), 14 tRNAs in plants and 23 in humans, 30 to 50 polypeptides in plants and 13 in humans. (Nass and Nass, 1962; Luck and Reich, 1964)

mitochondrial genetic code: It is similar to universal genetic code with some deviations due to RNA editing. UGA codes for tryptophane (stop codon in standard code), AUA for methionine (isoleucine in standard code), AGA, AGG for termination codons (arginine in standard code), and CGC for tryptophane (arginine in standard code).

mitochondrial plasmid: In the mitochondria of several higher plants and in some fungi, any of the small plasmid-like DNA molecules (minicircles and minilinears). (Pring et al., 1977)

mitochondrial replication: The formation of new mitochondria by division of existing ones. Divisions are accomplished by invagination of the inner membrane.

mitochondrial ribosome: A ribosome associated with the mitochondrion having a sedimentation constant of 70S.

mitochondrion (pl: mitochondria): A small cytoplasmic organelle where cellular respiration occurs. (Benda, 1898)

mitogens: Substances that provoke cell division (mitosis).

mitomycin C: An antibiotic, derived from *Streptomyces caespitosus,* that inhibits DNA synthesis and mitosis by cross-linking DNA, being particularly potent against Gram-positive, Gram-negative, and acid-fast bacilli. It is used in immunology to prevent immunologically reactive cells from dividing, and its use has also been suggested for cancer therapy.

mitosis: The series of events that result in the division of a single cell into two daughter cells. (Flemming, 1882)

mitotic analysis: Examination of the mitotic division cycle. The purpose might be to count the number of chromosomes at metaphase or to calculate the rate of cell division, etc.

mitotic apparatus: *See* SPINDLE.

mitotic crossover: A crossover resulting from the pairing of homologs in a mitotic diploid.

mitotic death: Cells fatally damaged by ionizing radiation may not die until the next mitosis, at which point the radiation damage to the DNA becomes evident, particularly when there is fragmentation of chromosomes.

mitotic index (MI): The fraction of a sample of cells that are undergoing mitosis. MI is commonly expressed as values below 1, so that 0.1 represents 10 percent of cells in mitosis. Percentages of cells in mitosis can vary greatly, from more than 10 percent in tissue culture to much less than 1 percent in relatively slowly growing tissue in vivo. (Minot, 1908)

mitotic nondisjunction: Occurs when sister chromatids fail to migrate to opposite poles of the cell during mitotic anaphase. The result is daughter cells with hyperploid and hypoploid chromosome counts.

mitotic recombination: Crossover between homologous chromosomes during mitosis, which leads to the segregation of heterozygous alleles.

mitotic-meiotic switch: Refers to critical biochemical events in the cell that lead to a shift from their mitotic to meiotic behavior.

mixed families: Groups of four codons sharing their first two bases and coding for more than one amino acid.

mobile dispersed genetic element: *See* TRANSPOSON. (Ilyin et al., 1980)

model: (1) The organism that is mimicked by another organism; (2) a mathematical description of a biological phenomenon.

model building: An experimental approach in which possible structures of biological molecules are assessed by building scale models of them.

moderately repetitive sequences: When the number of copies of a sequence in eukaryotes is 10^3 to 10^5, these sequences are interspersed with unique sequences and are also called *dispersed repeats*.

modification enzyme: An enzyme that recognizes the same nucleotide sequence as a corresponding restriction endonuclease and methylates certain bases within the sequence, thus conferring protection from the endonuclease activity.

modification of DNA/RNA: Includes all changes made to the nucleotides after their initial incorporation into the polynucleotide chain.

modified bases: All those except the usual four from which DNA (T, C, A, G) or RNA (U, C, A, G) are synthesized; they result from postsynthetic changes in the nucleic acid.

modified single cross: The progeny of a cross between a single cross, derived from two related inbred lines, and an original nucleus.

modifier (modifying) gene: A gene that affects or modifies the expression of another gene. (Bridges, 1919)

modulating codon: DNA coding triplets (codons) that code for rare transfer RNA (tRNA) molecules. Since the tRNA that carries the appropriate amino acid to the mRNA is in short supply in the cell, the presence of such codons tends to slow down translation of an mRNA containing them. (Ames and Hartman, 1963)

molecular biology: The branch of biology devoted to the study of the molecular nature of the gene and its biochemical reactions, such as transcription and translation. (Astbury, Beighton, and Weibull, 1955)

molecular clock hypothesis: A hypothesis postulating that the rate of molecular evolution is approximately constant over time among different evolutionary lineages and reflects the divergence time between taxa. (Zukerbandl and Pauling, 1965)

molecular cloning: The multiplication of DNA sequences usually involving the isolation of appropriate DNA fragments and their *in vitro* insertion into a restriction site of a cloning vector capable of replication when introduced into an appropriate host. This process requires (a) DNA of interest, (b) a cloning vector, and (c) a prokaryotic or eukaryotic cell to serve as a host.

molecular drive: A molecular mechanism that explains evolution in multiple gene families. It includes unequal recombination, transposition, and gene conversion. (Dover, 1982)

molecular genetics: The branch of genetics that attempts to explain various genetic phenomena at molecular level.

molecular hybridization: The formation of duplexes of complementary RNA or DNA strands under controlled conditions *in vitro,* thus, single-stranded nucleic acids become double-stranded. The progress of the reaction is monitored using methods that can detect this change, such as spectrophotometry.

molecular imprinting: The phenomenon in which there is differential expression of a gene depending on whether it was maternally or paternally inherited.

molecular marker: A sequence that can be derived from any molecular data which provides a screenable polymorphism between two organisms that are to be compared. Desirable properties of molecular markers are: (a) a high level of polymorphism, (b) codominant inheritance, (c) unambiguous designation of alleles, (d) frequent occurrence in the genome, (e) even distribution throughout the genome, (f) selectively neutral behavior, (g) easy access, (h) easy and fast assay, (i) higher reproducibility, (j) easy exchange of data between laboratories, and (k) development at reasonable cost.

molecular tinkering: An evolutionary process responsible for production of a protein with new properties, by joining together pieces of several different genes.

molecular zippers: These are the regulatory proteins which reveal the motifs which are composed primarily of amino acids that may not make direct contact with DNA. These motifs form three-dimensional scaffolds that steer the particular amino acid side chain of a regulatory protein into the grooves of double helical DNA where they can interact directly with DNA bases. These zippers may be of several types: steroid receptors, helix-turn-helix motifs, acid blobs, amphipathetic helix-loop-helix motifs, zinc finger motifs, and leucine zippers.

monintron gene: The genes that contain only one intron. Transcripts of such genes are not used for translation.

monocentric: Having a single centromere.

monocistronic mRNA: An RNA molecule, mostly in eukaryotes, that contains information from only one cistron.

monoecious: Individuals producing both sperm and egg. (Darwin, 1876)

monogenic inheritance: Pattern of inheritance observation in a cross between two individuals identically heterozygous at one gene pair. It gives a 3:1 ratio in F_2.

monohybrid: The offspring of two homozygous parents that differs by only one gene locus where only one such locus is under consideration. (DeVries, 1901)

monohybrid cross: A sexual cross in which the inheritance of just a single pair of alleles is followed (e.g., tall × tall).

monohybrid heterosis: Hybrid vigor due to heterozygosity at a single gene locus.

monoisodisomic: An individual deficient in one chromosome but having an isochromosome for one of the arms of the missing chromosome. Also called *haplo-triplodisomic* by Khush and Rick (1967). (Kimber and Sears, 1968)

monoisosomic: An individual that is missing one chromosome pair but has an isochromosome for one arm of the missing pair. (Kimber and Sears, 1968)

monomorphic: A gene having a single form or trait in a population.

monophyletic group: In cladistic analysis, a group in which any species belonging to the group is more closely related to any other species also in the group than to any species that does not belong, by virtue of having at least one shared characteristic that defines the group.

monophyletic species: A group of species derived from a single ancestral population.

monoploid: An individual with a single complete set of chromosomes; also the fundamental number of chromosomes comprising a single set. *Compare* HAPLOID. (Langlet, 1927)

monosome: A chromosome that has no homologue present; a haploid chromosome in an otherwise normal diploid individual. (Haselkorn and Fried, 1964)

monosomic: An individual lacking one chromosome of a set (2n – 1). (Blakeslee, 1921)

monosomic alien addition lines (MAALs): A set of lines each one of which contains a different extra chromosome from a related species.

monosomy: A chromosome aberration in which only one of a given kind of chromosome is present instead of the normal two characteristic of a typically diploid organism.

monotelodisomic: An individual deficient in one chromosome arm. (Kimber and Sears, 1968)

monotelomonoisosomic: An individual who has one chromosome pair missing but has a telocentric chromosome for one arm of the missing pair and an isochromosome for the other arm. (Kimber and Sears, 1968)

monotelosomic: An individual deficient in one entire chromosome and one arm of the other homolog.

monotypic species: A species that consists of a uniform population devoid of differentiation and semi-isolated species.

Morgan unit: The length of chromosome in which an average of one recombination event occurs each time a gamete is formed. It equals a crossover value of 100 percent. (Haldane, 1919)

morph: Any relatively common heritable variation; any of the variant types of a polymorphic species. (Huxley, 1957)

morphogen: A factor that induces development of particular cell types in a manner that depends on its concentration. (Turing, 1952; Gierer, 1977)

morphogenesis: (1) The evolution of a structure from an undifferentiated to a differentiated state; (2) the process of growth and development of differentiated cell types from genetically identical and morphologically alike cells.

morphological mutations: Mutations that affect the morphological phenotype of an individual.

morphological species concept: Organisms are classified in the same species if they appear similar in form.

morphology: Study of the form of an organism—developmental history of visible structures and the comparative relation of similar structures in different organisms.

morphospecies: Species described wholly in terms of morphological characteristics.

mosaic: An individual, part of whose body is composed of tissue genetically different from another part.

mosaic gene: *See* CHIMERIC GENE.

mother cell: A cell with a diploid nucleus that, by meiosis, produces four haploid nuclei.

mRNA: *See* MESSENGER RNA.

MS2: Bacteriophage with an RNA genome and from which an RNA replicase has been isolated.

mtDNA: DNA that occurs as a normal component within mitochondria. *See* MITOCHONDRIAL (mt) DNA.

multiallelic variation: Genes or DNA segments showing a high level of polymorphic variation, exemplified by a variable number of tandem repeats (VNTRs) and microsatellites. *See* VARIABLE NUMBER OF TANDEM REPEAT (VNTR) LOCI and MICROSATELLITES.

multicloning site (MCS): A short DNA sequence, found in most vectors in common use, that contains many closely spaced, restriction-enzyme cleavage sites. Also known as *polylinker*.

multicopy plasmids: Plasmids that are present in cells in numbers exceeding one per cellular genome copy; e.g., pBR322 is a multicopy plasmid, and there are usually 50 pBR322 copies per *E. coli* genome. *See* RELAXED CONTROL.

multifactorial (polygenic) trait: A trait whose phenotypic expression is influenced by the cumulative effects of many genes.

multiforked chromosome: This results when a bacterial chromosome has more than one replication fork because a second initiation has occurred before the first cycle of replication has been completed.

multigene family: A group of genes, possibly although not always clustered, that are related either in nucleotide sequence or in terms of function. Two basic patterns of multiple-gene families have emerged:

(a) linked clusters of related genes in which individual genes are separated by long regions of noncoding DNA, and (b) dispersed families in which individual genes are scattered over widely separated chromosome locations. Multiple-gene families may include truncated genes, pseudogenes, or processed genes. (Brown, Wensink, and Jordan, 1972; Hood, 1972)

multigenic inheritance: Inheritance determined by several genes with cumulative effect, e.g., ear length in corn.

multihybrid: An organism heterozygous at numerous loci.

multiline cultivar: A cultivar developed by compositing in varying proportions many isolines, each differing in its resistance to different races of pathogen, belonging to a particular, highly adapted cultivar in self-pollinated crops. These cultivars slow down the onset of an epidemic.

multiline variety: A composite of isolines.

multintron gene: A gene that contains more than one intron. Transcripts of such genes are translated into proteins.

multipartite genome organization: The main chromosome having all the genetic information is called the master chromosome. *See* MASTER CHROMOSOME. (Palmer and Shields, 1984)

multipathotype tests: Seedlings of host plants are tested individually in the greenhouse for resistance against a range of pathotypes. These tests permit postulation of some resistance genes in the host. Since these tests are usually performed on seedlings, important adult plant resistance genes may remain unidentified.

multiple alleles: The different alternative states of a gene with more than two alleles. The existence of several known alleles of a gene is known as *multiple allelism*.

multiple convergence: When the desirable characters are distributed among a number of genotypes, a complex, cyclic, crossing schedule is employed, e.g., [(A × B) × (C × D)] × [(E × F) × (G × H)] to form a complex hybrid. This approach leads to convergence of several small streams of germplasm into one large stream for inbred improvement and their use in crosses.

multiple factor hypothesis: A hypothesis that explains quantitative inheritance on the basis of action and segregation of a number of allelic pairs having duplicate and accumulative effects without complete dominance; the same as quantitative inheritance.

multiple genes: Two or more independent pairs of genes that produce complementary or cumulative effects upon a single character of the phenotype.

multiplication rule: If an event occurs in n_1 ways, and a totally independent event can occur in n_2 ways, the number of ways both events can occur at the same time is $n_1 \times n_2$.

multisite mutant allele: A mutant differing from its wild-type form at two or more sites.

multivalent: An association of more than two chromosomes whose homologous regions are synapsed by pairs.

mutability: The ability to change.

mutable gene: A gene that exhibits a higher mutation rate than others.

mutable site: Sites along the chromosome at which mutations can occur. Genetic experiments tell us that each mutable cell site can exist in several alternative forms.

mutafacient: A gene or genetic element that determines or increases the chance of mutation of another gene or genetic element.

mutagen: A chemical or physical agent able to cause a mutation in a DNA molecule.

mutagenesis: The experimental treatment of a group of cells or organisms with a mutagen in order to induce mutations.

mutagenic DNA alterations: Any of those primary DNA alterations that do not prevent DNA replication and may give rise to gene mutations. They may arise due to mistakes in base pairing or may result from mistakes by the DNA polymerase. (Freese, Phaese, and Freese, 1969)

mutagenic effectiveness: The ratio of factor mutations to dose; i.e., the frequency of a given mutation to occur at a given dose of mutagen employed.

mutagenic efficiency: The ratio of factor mutations to biological damage; i.e., desirable changes free from associated undesirable changes on mutagenesis.

mutant: A gene, cell, or organism with an abnormal genetic constitution that may result in a variant phenotype.

mutant allele: An allele differing from the allele found in the standard or wild-type form.

mutant hunt: The process of accumulating different mutants showing abnormalities in a certain structure or function as a preparation for mutational dissection of that function.

mutant site: The damaged or altered area within a mutated gene.

mutasome: In SOS response, a complex of activated Umu D protein with Umu C and RecA proteins and DNA polymerase III. The mutasome is located at the lesion and may permit the polymerase to replicate past it.

mutation: An alteration in the nucleotide sequence of a DNA molecule. (DeVries, 1901)

mutation breeding: The use of mutagens to develop variants that can increase agricultural yield.

mutation event: The actual occurrence of a mutation in time and space.

mutation frequency: The frequency at which a particular mutant is found in the population.

mutation pressure: A constant mutation rate that adds mutant genes to a population. (Wright, 1921)

mutation rate: The proportion of mutants per cell division in bacteria or single-celled organisms or the proportion of mutations per gamete in higher organisms. (Muller, 1932)

mutation spectrum: A term used to indicate that an organism may produce different kinds of mutants due to the differential action of various mutagens.

mutation theory of evolution: This theory advocates that a single mutation can lead to radical changes in an organism which can transform one species to another. *See* DEVRIESISM. (DeVries, 1901)

mutation trend: A series of slight gene mutations in the same direction resulting in gradual intensification of a character change. (Pincher, 1946)

mutational dissection: The study of the components of a biological function through a study of mutations affecting that function.

mutational equilibrium: The equilibrium point at which the forward mutations balance the reverse mutations so that the allelic frequencies do not change as a consequence of those mutations.

mutational load: Genetic load caused by spontaneous gene mutation. (Muller, 1950)

mutator gene: A gene that enhances the mutation rate of another gene in the same genome.

mutator mutations: Mutations of DNA polymerase that increase the overall mutation rate of a cell or of an organism.

muton: The smallest segment of DNA or subunit of a cistron that can be changed and thereby bring about a mutation; can be as small as one nucleotide pair. (Benzer, 1957)

mutual translocation: Reciprocal transfer, or crossing-over, between the terminal portions of two nonhomologous chromosomes.

n, 2n: The genetic (haploid) and zygotic or somatic (diploid) chromosome numbers respectively. (Blakeslee, 1921)

N banding: Minor modification of C banding technique where specialized heterochromatin stains more densely within C bands. N banding has been used successfully in identification of wheat chromosomes.

nanometer (nm): A unit of length or diameter equal to 1×10^{-9} meter. A synonym is millimicron (mμ).

narrow heritability: *See* HERITABILITY IN NARROW SENSE.

natural selection: The process in nature whereby one genotype leaves more offspring than another because of superior life-history attributes, such as survival or fecundity. Darwin proposed natural selection as a mechanism of evolution.

nearest-neighbor analysis: A technique of transferring radioactive atoms between adjacent nucleotides in DNA used to demonstrate that the two strands of DNA run in opposite directions.

negative complementation: The complementation that occurs when interallelic complementation allows a mutant subunit to suppress the activity of a wild-type subunit in a multimeric protein.

negative control: The repression of an operator site by a regulatory protein that is produced by a regulator site.

negative heteropyknosis: Deficient charging of heterochromatin with nucleic acid in meiotic and premeiotic divisions.

negative interference: The phenomenon whereby a crossover in a particular region enhances the occurrence of other apparent crossovers. *Compare* POSITIVE INTERFERENCE.

negative regulation: Negative feedback in biological systems mediated by allosteric regulatory enzymes.

negative regulators: The molecules function by switching off transcription or translation.

negative supercoiling: The twisting of a duplex of DNA in space in reverse direction to the turns of the strands in the double helix.

N-end rule: The life span of a protein is determined by its amino-terminal (N-terminal) amino acid.

neocentric activity: The association between a site of the chromosome other than the centromere and fibers of the spindle apparatus.

neo-Darwinism: A term that refers to the merger of classical Darwinian evolution with population genetics and thus leading to the synthesis of the modern theory of evolution. *See* SYNTHETIC THEORY OF EVOLUTION.

neolithic revolution: This occurred when man first became settled in one place and depended upon domesticated animals and plants; this period of cultural development, during which agriculture originated, is known as the neolithic revolution.

neomorph: An allele having an effect apparently unrelated to that of the wild-type allele. (Muller, 1932)

neutral alleles: The alleles whose differential contribution to fitness is so small that their frequencies change more due to generations than to natural selection.

neutral gene hypothesis: The hypothesis that most genetic variation in natural populations is not maintained by selection.

neutral mutations: These are same-sense or missense mutations having no harmful effect on the function of the gene.

neutral substitutions: Amino acid substitutions in a protein that do not affect the activity of the molecule.

neutrality: *See* SELECTIVE NEUTRALITY.

neutrality theory of protein evolution: The rates of amino acid replacements in proteins and nucleotide substitutions in DNA during evolution may be approximately constant because the vast majority of such changes are selectively neutral.

Newcommer's fluid: A fixation used in chromosome analysis composed at the ratio of 6 parts isopropyl alcohol : 3 parts propionic acid : 1 part petroleum ether : 1 part acetone : 1 part dioxane (with or without ferric acetate).

niche: The totality of environmental factors into which a species (or other taxon) fits; the outward projection of the needs of an organism—its specific way of utilizing its environment. (Grinnell, 1917)

nick: In duplex DNA, nick is the absence of a phosphodiester bond between two adjacent nucleotides on one strand.

nick translation. The ability of *E. coli* DNA polymerase I to use a nick as a starting point from which one strand of a duplex DNA can be degraded and replaced by resynthesis of new material. Nick translation is used to introduce radioactivity labeled nucleotides into DNA *in vitro*. (Kelly, 1970)

nickase: *See* DNA GYRASE.

nick-closing enzyme: A form of DNA polymerase that can restore base sequences. A nick is introduced adjacent to an incorrect base. The base is then removed and the correct base inserted.

nicking: Nuclease action to sever the sugar-phosphate backbone in one DNA strand at one specific site.

nicking-closing enzyme: An enzyme involved in the normal DNA replication. The enzyme in *E. coli* is a monomer (Mr = 110,000) that uses negatively supercoiled DNA as a template. Its function is to relax supercoiling stress that would otherwise build up in the DNA during unwinding for replication. Also known as *topoisomerase I*. *See* NEGATIVE SUPERCOILING. (Wang, 1971)

nif gene: The genetic notation for the genes involved in nitrogen fixation. The nif operon is a complex array of seventeen genes. The proteins encoded by the nif genes will fix atmospheric nitrogen (N_2) into ammonia (NH_4^+) and nitrate (NO_3^-). Many soil bacteria will fix nitrogen, and there is much interest in manipulation of nif genes from bacteria to allow plants to fix nitrogen. The formation of nitrogen compounds (NO_2^-, NO_3^-) from free atmospheric nitrogen (N_2) is known as *nitrogen fixation*. *Nitrogenase* is an enzyme complex that catalyzes the reduction of atmospheric nitrogen (dinitrogen) to ammonia in nitrogen-fixing organisms. The enzyme has two major components, each of which has two or four subunits. The large component contains molybdenum, nonhaem iron, and sulphur; the smaller contains iron and sulphur. The enzyme, which is strictly anaerobic and is easily inactivated by molecular oxygen, also acts as a hydrogenase. Hydrogen production is a wasteful reaction, and those species with a low level of hydrogen production are more efficient nitrogen fixers. *Rhizobium* bacteria are found in special nodules on the roots of the plants and within the bacteria where nitrogen fixation takes place. The plants are provided with NH_2, and the bacteria obtain carbon compounds for growth from the plant. Many, if not all, Rhizobia contain one or more large plasmids, some of which give the bacterium the ability to colonize a particular species of plant.

NIH guidelines: These are recommended procedures for the conduct of recombinant DNA experiments to which all National Institute of Health (NIH) grantholders must adhere.

nitrogenous base: One of the purine or pyrimidine compounds that form part of the molecular structure of a nucleotide.

nitrous acid (HNO_2) as a mutagen: An acid that acts on cells as a mutagen by converting the purine and pyrimidine base amino groups to hydroxyl groups.

no selection: *See* ZERO SELECTION.

nonambiguous code: The nature of a genetic code where one codon codes for only one amino acid.

nonautonomous controlling elements: Defective transposons that can transpose only when assisted by an autonomous controlling element of the same type.

noncoding DNA: DNA that is in the nonsense strand (negative strand) of the DNA duplex or is out with the coding sequence of any gene. It may or may not be taken to include intron sequences. The centromeric region of DNA is noncoding.

noncoding DNA strand: By convention, the base sequence for a transcribed region on this strand is that of the messenger RNA transcribed from it. Its advantage is that the amino acid encoding codons and the start and stop codons are in their standard forms and thus easily recognizable.

non-Darwinian evolution: A theory of evolution that considers natural selection as incompetent to account for survival of the fittest. Saltation hypothesis, punctuation equilibrium model, and evolution by random walk are in contrast to the concept of Darwinian evolution. *See* SALTATION HYPOTHESIS, PUNCTUATION EQUILIBRIUM MODEL, and EVOLUTION BY RANDOM WALK. (Harris, 1968; King and Jukes, 1969)

nondisjunction: The failure of homologous chromosomes to separate at anaphase I of meiosis. (Bridges, 1913)

non-DNA repair: These are intrinsic features of the organism, such as structure of the code, intergenic and intragenic suppression, and diploidy, in addition to normal repair mechanisms.

nongenetic RNA: The RNA that is found in association with DNA and does not act as genetic material.

nonhistone proteins: The proteins remaining in chromatin after the histones are removed. The scaffold structure is made of nonhistone proteins. (Mayfield and Ellison, 1975)

nonhomologous recombination: Recombination between two double-stranded DNA molecules that have little or no nucleotide sequence similarity.

non-Mendelian ratio: An unusual ratio of progeny phenotypes that does not reflect the simple operation of Mendel's laws; for example, a mutant:wild ratio of 3:5, 5:3, 6:2, or 2:6 in tetrads indicates that gene conversion has occurred.

nonoverlapping codon: The nature of a genetic code where one nucleotide is part of only one codon.

nonparentals: *See* RECOMBINANT.

nonpermissive conditions: The conditions that do not allow lethal mutants to survive.

nonpreference: Plant resistance to insects through suppression of feeding or oviposition.

nonrandom assortment: An assortment or segregation of genes created by a process that is not random; e.g., linked genes show nonrandom assortment.

nonrandom mating: Refers to selective mating when specific phenotypes show a preference for a particular phenotype; e.g., people tend to marry people of their own race.

nonreciprocal recombination: Recombination in which homologous chromosomes do not undergo reciprocal exchange of genetic material. Only one chromosome may be recombinant, the other never forming.

nonrecombinants: In mapping studies, offspring that have alleles arranged as in the original parents.

nonrecurrent parent: A parent not involved in a back cross. Also known as *donor parent*. *See* RECURRENT PARENT.

nonrepetitive DNA: The DNA that shows reassociation kinetics expected of unique sequences.

nonsense codon: One of the mRNA sequences (UAA, UAG, UGA) that signals the termination of translation. (Brenner et al., 1961)

nonsense intergenic suppressor: A gene coding for a mutant tRNA able to respond to one or more of the nonsense codons. (Capecchi and Gussin, 1965)

nonsense mutation: An alteration in a nucleotide sequence that converts triplet coding for an amino acid into a termination codon.

nonsister chromatids: The chromatids that are derived from partner chromosomes at pachytene, or from any distinct pair of chromosomes at mitosis.

nontranscribed spacer: The region between transcription units in a tandem gene cluster.

nonviral retroelement: A mobile DNA element that transposes via an RNA intermediate that is copied into DNA by reverse transcription. Some nonviral retroelements are similar to retroviruses but lack genes for capsid proteins.

nopaline: A rare amino acid derivative that is produced by a certain type of crown-gall tissue. The genes responsible for the synthesis of nopaline are part of the T-DNA from a Ti plasmid.

NOR: *See* NUCLEOLAR ORGANIZER REGION.

norientation: Descriptive of the situation in which the vector and the inserted fragment of DNA have the same orientation.

norm of reaction: The pattern of phenotypes produced by a given genotype under different environmental conditions. (Woltereck, 1909)

normal curve: A bell-shaped curve that represents the distribution of individuals segregating for a quantitative character, such as height, yield, intelligence, etc. Usually, the curve is humped in the midway between two extremes, indicating that the majority of individuals in the population are grouped in this class called a mode or model class. There may be two modes. The mode may be nearer to one end of the curve or the other, in which case it is called a *skewed curve*.

normal distribution: Any of a family of bell-shaped frequency curves whose relative position and shape is defined on the basis of the mean and standard deviation.

normalizing selection: The removal by selection of all genes that produce deviations from the normal (= average) phenotype of a population.

Northern blotting: A technique used for transferring RNA from an agarose gel to a nitrocellulose filter on which it can be hybridized to a complementary, radiolabeled, single-stranded DNA or RNA probe.

N-terminus: The amino ($-NH_2$) end of a peptide chain, by convention written as the left end of the structural formula.

Nu body: The subunits of chromatin produced during chromosome coiling in eukaryotes with roughly spherical shape, which is composed of a core octamer of histones and approximately 140 nucleotide pairs of DNA. Also known as *nucleosome*. (Olins and Olins, 1973)

nuclear genome: In eukaryotes, the genetic information encoded in the nuclear DNA, as opposed to the organellar genomes (mitochondrial and chloroplast genomes).

nuclear sap: The fluid that is lost by the chromosomes as they contract during prophase and that fills the space of the nucleus.

nuclear system of heredity: The normal, Mendelian system in which inheritance is controlled by genes located on the chromosomes, as opposed to the corpuscular (= plastid) system and the cytoplasmic (= molecular) system.

nuclear transplantation: The technique of placing a nucleus from another source into an enucleated cell.

nuclease: An enzyme that degrades a nucleic acid molecule.

nuclease hypersensitive site: The region of eukaryotic chromosomes that is specifically vulnerable to nuclease attack because it is not wrapped, as with nucleosomes. (Varshavsky, Sundine, and Bohn, 1978)

nuclease-free reagent: This is a chemical used to extract or purify DNA or RNA that has been tested to ensure very low levels of ultraviolet-absorbent material and enzymes capable of denaturing or hydrolyzing nucleic acids (RNA and DNA).

nucleic acid: Originally, the acidic chemical compound isolated from the nuclei of eukaryotic cells. Now, the polymeric molecules comprising nucleotide monomers DNA and RNA. (Altmann, 1889)

nucleic acid base: A component of nucleic acid molecules including adenine, thymine, uracil, guanine, and cytosine.

nucleic acid hybridization: The formation of a double-stranded molecule by base pairing between complementary polynucleotides.

nuclein: The nucleoprotein complex isolated by Meischer (1869).

nucleocytoplasmic interaction: Activity of nuclear genes during development is limited by properties of the cytoplasm. Phenotype is regulated but not determined by cytoplasm. Genes determine a cell's potential; the cytoplasm determines whether or not that potential will be reached.

nucleoid: The fibrillar structure in prokaryotes where the chromosome is located. (Piekarsky, 1937)

nucleolar organizer region: The chromosomal region located around the secondary constriction around which the nucleolus forms during interphase, prophase, and telophase; the site of tandem repeats of the major rRNA genes. (McClintock, 1934)

nucleolus: The region of the nucleus in which rRNA transcription occurs. (Valentine, 1836)

nucleolytic: A reaction involving hydrolysis of a phosphodiester bond in a nucleic acid.

nucleoplasm: A general name for the complex mixture of molecules that comprises the ground substance of the nucleus of a living cell.

nucleoprotein: The substance of eukaryotic chromosomes consisting of proteins and nucleic acids.

nucleoside: A chemical compound comprising a purine or pyrimidine base attached to a 5-carbon sugar.

nucleosome: The structure comprised of histone proteins and DNA that is the basic organizational unit in chromatin structure. *See* NU BODY. (Oudet, Gross-Bellard, and Chambon, 1975)

nucleosome bead: An octamer of histones plus 146 base pairs of DNA wrapped around the octamer in two turns. Also known as *core particle*.

nucleotide: A chemical compound comprising a purine or pyrimidine base attached to 5-carbon sugar, to which a mono-, di-, or triphosphate is also attached. The monomeric unit of DNA and RNA. (Levene and Bass, 1931)

nucleotide diversity: Variations in a nucleotide sequence of a particular DNA/RNA fragment or gene.

nucleotide pair: A pair of nucleotides (one in each strand of DNA) that are joined by hydrogen bonds.

nucleotide pair substitution: The replacement of a specific nucleotide pair by a different pair, often mutagenic. More commonly known as *base pair substitution.*

nucleus: The membrane-bound structure of a eukaryotic cell within which the chromosomes are contained. (Brown, 1831)

null allele: An allele whose protein product shows no histochemically detectable activity.

null hypothesis: The hypothesis from which the expectations are predicted for the purpose of testing significance.

null mutation: A mutation that completely eliminates the function of a gene, usually because it has been physically synthesized.

nullihaploid: *See* NULLISOMIC HAPLOID.

nulliplex: The condition in which a polyploid is recessive in all chromosomes in respect to a particular gene. Simplex denotes recessiveness at all loci except one, duplex two, triplex three, quadriplex four, etc. (Belling and Blakeslee, 1926)

nullisomic: An aneuploid in which both members of one particular pair of homologous chromosomes are missing from the chromosome complement. (Blakeslee, 1921; Sears, 1953)

nullisomic haploid: A haploid individual lacking one chromosome ($x - 1$, $2x - 1$, $3x - 1$). It is derived from disomic alien addition line. (Riley and Chapman, 1958)

nullisomic-tetrasomics: Lines that lack one chromosome pair but in which the loss is compensated for by the duplication of another related pair of chromosomes.

nurse culture: In the culture of plant cells, the growth of a cell or cells on a contiguous culture of different origin that, in turn, is in contact with the tissue culture medium. The cultured cell or tissue may be separated from the feeder layer by a porous matrix such as filter paper or membranous filters. *See* FEEDER LAYER.

nutrient medium: A solid or liquid combination of nutrients and water, usually including several salts, a carbohydrate (e.g., sucrose), and vitamins. Such a medium is often referred to as a basal medium and may be supplemented with growth hormones and occasionally, with other defined and undefined substances.

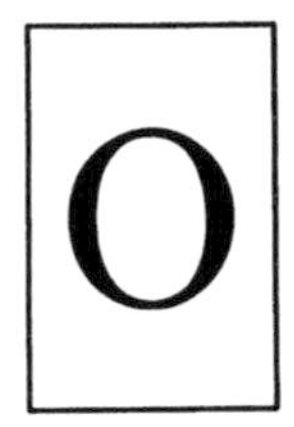

obligate pathogen: A disease-causing microorganism that requires a living host in which to grow and reproduce. *Compare* FACULTATIVE PATHOGEN.

ochre codon: The nonsense codon UAA. One of the three nonsense codons that cause termination of protein synthesis.

ochre mutation: Any change in DNA that creates a UAA codon that causes termination of protein synthesis. (Brenner and Beckwith, 1965)

ochre suppressor: A gene coding for a mutant tRNA able to respond to the UAA codon to allow continuation of protein synthesis.

octopine: A rare amino acid derivative that is produced by a certain type of crown-gall tissue. The genes responsible for the synthesis of octopine are part of the T-DNA from a Ti plasmid.

octoploid: An individual with eight genomes or monoploid sets of chromosomes. Octoploids customarily form chromosome pairs at meiosis and not "n" sets of eight.

Okazaki fragment: One of the short segments of RNA-primed DNA that is synthesized during replication of the lagging strand of the double helix. *See* DISCONTINUOUS REPLICATION.

oligodeoxynucleotide-directed mutagenesis: An *in vitro* mutagenesis technique that involves the use of a synthetic oligonucleotide to introduce a predetermined nucleotide alteration into the gene to be mutated. (Razin et al., 1978)

oligogenes: One or a few genes governing the same qualitative character. *See* MAJOR GENE.

oligogenic inheritance: Inheritance of oligogenic or qualitative characters. *See* QUALITATIVE INHERITANCE. (Mather, 1941)

oligonucleotide: A linear sequence of a few (generally not over ten) nucleotides.

oligonucleotide fingerprinting: The first microsatellite-based marker technique was derivative of RFLP (restriction fragment length polymorphism) analysis using microsatellite-complementary oligonucleotides as probes. Oligonucleotide printing comprises several steps: (a) isolation of high-molecular-weight genomic DNA, (b) complete digestion of the genomic DNA, (c) electrophoretic separation of the restriction fragments, (d) denaturation and immobilization of the separated DNA fragments, (e) hybridization of the dried gel to labeled microsatellite-complementary oligonucleotide probes, and (f) detection of hybridizing fragments (i.e., fingerprints) by autoradiography or by chemiluminescence and their documentation by photographing.

oligonucleotide ligation assay (OLA): This assay is based on the ability of two dinucleotides to anneal immediately adjacent to each other on a complementary target DNA molecule.

oligonucleotide probe design: The use of oligonucleotides for gene amplification, as diagnostic probes or antisense-based drugs, can be optimized by minimizing the possibility of nonspecific hybridization. Oligonucleotides are also used in the blocking of translation-specific genes. (Mitsuhashi et al., 1994)

oligonucleotide therapy: The use of oligoribonucleotides and oligodeoxynucleotides as the therapeutic agents. Three basic strategies are: antisense inhibition of expression, triple helix formation, and protein epitope targeting.

oligopeptide: A peptide polymer containing somewhere between two and ten amino acids.

O-micron DNA: An alternative name for the 2 percent plasmid of *Saccharomyces cerevisiae*.

one gene-one enzyme hypothesis: One gene controls the synthesis of one enzyme. This hypothesis is valid for only those enzymes or proteins that are made up of only one type of polypeptide. (Beadle and Tatum, 1941)

one gene-one polypeptide hypothesis: One gene controls the synthesis of only one polypeptide. This hypothesis is valid for all the protein-encoding genes.

one gene-one primary function hypothesis: One gene performs one specific primary cellular function.

ontogeny: Growth (usually size increase) and development (both size increase and differentiation of traits) of the individual. (Haeckel, 1894)

oocyte, primary (2N): The diploid cell that will undergo meiosis to form an egg.

oocyte, secondary (N): The haploid cell following the first meiotic division that divides equationally to produce the egg (female gamete) and an additional polar body that disintegrates.

oogenesis: Egg formation.

oogonia: Cells in females that produce primary oocytes by mitosis.

opal codon: The nonsense codon UGA. One of the three nonsense codons that causes termination of protein synthesis.

opal suppressor: A gene which codes for an altered transfer RNA so that its anticodon can recognize the opal codon and thus, allows the continuation of protein synthesis. A suppressor of an opal mutation is a tRNA that is charged with the amino acid corresponding to the original codon or a neutral substitute. Some eukaryote cells normally synthesize opal suppressor tRNAs. The function of these is not clear and they usually do not prevent normal termination of protein synthesis at an opal codon.

open circle: One of the three conformations that a plasmid molecule may adopt. In an open circle, there has been either a single break or a number of staggered breaks made in the strands of the duplex. Such a molecule may not form supercoils and therefore, is referred to as an open or relaxed circle.

open continuous culture: A continuous culture is one in which inflow of fresh medium is balanced by outflow of a corresponding volume of culture. Cells are constantly washed out with the outflowing liquid. In a steady-state, the rate of cell washout equals the rate of formation of new cells in the systems.

open promoter complex: The complex formed between *E. coli* RNA polymerase and a promoter in which the double helix is partially unwound in readiness for the start of RNA synthesis.

open reading frame (ORF): A series of codons with an initiation codon at the 5′ end but no termination codon. Often considered synonymous with "gene" but more properly used to describe a DNA sequence that looks like a gene, but to which no function has been assigned.

operator: A nucleotide sequence element to which a repressor protein attaches in order to prevent transcription of a gene or operon. (Jacob and Monod, 1959)

operator-constitutive mutation (OC): A mutation in the operator that leads to constitutive synthesis of all the gene products of the operon, regardless of presence or absence of the substrate. The mutation affects activity of all the genes of the operon in *cis*-arrangement not in *trans*-arrangement.

operon: A system of cistrons, operator and promoter sites, by which a given genetically controlled, metabolic activity is regulated. (Jacob et al., 1960)

opine: The general name given to rare amino acids and sugar derivatives found in crown-gall tumors.

opine synthetase: The enzyme system responsible for the synthesis of opines of a number of different types. The genetic information coding for opine synthetase resides in the T-DNA region of the Ti plasmid. These plasmids confer virulence to strains of *Agrobacterium,* which are responsible for the crown-gall disease. The T-DNA is transferred to the plant cells, where it is stably integrated into the nuclear genome, and the opine synthetase gene is transcribed and translated.

opportunistic pathogens: Organisms that exist as a part of the normal body microflora but which may become pathogenic under certain conditions—e.g., when the normal antimicrobial defense mechanism of the host has been impaired.

opportunistic selection: Mass-pedigree selection where bulk handling is terminated whenever environment is conducive for some primary traits.

optical density (OD): The degree of absorbance of light by a solution as a function of the concentration of the solute. *See* ABSORBANCE.

orcein: Dye used in cytology, especially when dissolved in acetic acid, as with aceto-orcein, for the staining of squash preparations of chromosomes.

ordered tetrad: Linear sequences of the products of meiotic division as visualized in the row of eight ascospores of a fungal ascus. Crossing-over events can be studied by growing each spore separately and scoring for its expression of one allele at a particular locus.

ordinate: The vertical axis in a graph.

organ culture: The maintenance of growth of an organ primordia or the whole or parts of an organ *in vitro* in a way that may allow differentiation and preservation of the architecture and/or function.

organization effect: An interaction among adjacent loci owing to some features of organization of the chromosomes.

organogenesis: The process of differentiation of root and shoot from the somatic embryos. In plant tissue culture, a process of differentiation by which plant organs are formed *de novo* or from preexisting structures. In developmental biology, this term refers to differentiation of an organ system from stem or precursor cells.

orientation: The movement of centromeres so that they lie axially with respect to the spindle.

origin of replication (Ori): A site on a DNA molecule where unwinding begins in order for replication to occur.

origin of transfer (Ori-T): When gene transfer occurs through plasmids, one of the two strands of plasmid DNA is nicked at a site called Ori-T, and this linear strand moves into the recipient bacteria.

origin recognition complex (ORC): A multi-protein complex capable of ATP-dependent binding to yeast replication origins. (Bell and Stillman, 1992)

orphon genes: These are isolated individual genes found in isolated locations but related to members of a gene cluster. (Childs et al., 1981)

orthologous genes: Homologous genes that have become differentiated in different species that were derived from a common ancestral species. *Compare* PARALOGOUS GENES. (Finch et al., 1977)

outbreeding: The mating of genetically unrelated individuals.

out cross: Cross-pollination, usually by natural means, with a plant differing in genetic constitution.

outside markers: Loci on either side of another locus or specified region.

overdominance: Superiority of F_1 for one or more characters over both the parents. (Hull, 1946)

overdominance hypothesis: The theory that heterosis is caused by the Aa genotype being superior to either the AA or the aa parent. *See* DOMINANCE HYPOTHESIS.

overlapping code: The discredited idea that the genetic code consists of base pairs that participate in more than one codon.

overlapping deletion: Compound deletion occurring as two separate events in which one overlaps another.

overlapping genes: In the strictest sense, defined as a single nucleotide sequence coding for more than one polypeptide. Such an arrangement of genes was first discovered by Barrel, Air, and Hutchison (1976) in phage ϕX174. Later, these genes were discovered in the genomes of bacteriophages, animal viruses, and mitochondria, as well as in bacteria.

overlapping inversion: A situation in which a second inversion overlaps a previous, existing inverted sequence.

overwinding: Overwinding of DNA is caused by positive supercoiling, which applies further tension in the direction of the winding of the two strands about each other in the duplex.

p: (1) Relative frequency of a dominant gene; (2) short arm of a chromosome.

P: The probability with which a certain event may occur if the experiment is repeated under identical conditions. p = 85 means that if the experiment is repeated, 85 out of 100 times the chances are that the same results would be obtained.

P I, P II: Symbols for the prophase of the first (P I) and second (P II) meiotic divisions.

P_L, P_R: Promoter leftward and promoter rightward are two strong promoters from bacteriophage λ. Transcription from both these promoters is under the control of the cl gene product.

P_1, P_2, P_3, etc: First, second, third, etc. generations from a parent. Also used to designate different parents used in making a hybrid or series of hybrids. (Bateson and Saunders, 1902)

pachytene: A substage of prophase I of meiosis I during which homologous chromosomes are completely synapsed and undergo crossing-over. (Winiwarter, 1900)

packing ratio: The ratio of the length of DNA to the unit length of the fibre containing it. (Du Praw, 1970)

pairing of chromosomes: *See* SYNAPSIS.

pairing segments: The euchromatic regions of chromosomes that synapse with a homolog in the first meiotic prophase. (Darlington, 1931)

palindrome: A sequence of DNA base pairs that reads the same on the complementary strands. For example, 5′GAATTC3′ on one strand, and 5′CTTAAG3′ on the other strand. Palindrome sequences are recognized by restriction endonucleases. (Wilson and Thomas, 1974)

palindromic sequence: A bilaterally symmetrical DNA sequence that, therefore, reads the same in both directions. The target sites of most restriction enzymes are palindromic sequences; e.g.,

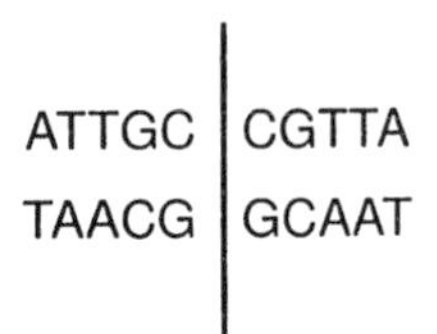

Palindromic sequences can form a cruciform structure due to intrastrand base pairing of the self-complementary sequence.

pan-editing: A spectacular type of RNA editing that in some cases accounts for 50 percent or more of the mature sequence of mRNA.

Pan-editing is a very primitive mechanism. (Hajduk, Harris, and Polland, 1993; Maslov, 1994)

pangene: The particle hypothesized by Darwin in his *pangenesis hypothesis* to explain variation and evolution.

paper chromatography: A chromatographic method using absorbent paper by which minute amounts of material can be analyzed. A paper strip with a drop of test material at the bottom is dipped into the carrier liquid (solvent) and removed when the solvent front almost reaches the top of the strip. Two-dimensional chromatograms can be produced using square paper and two different solvents. The paper is removed from the first carrier liquid, turned at right angles, and dipped into the second. This gives a two-dimensional map of the constituents of the test drop. The identity of the constituents may be found by measuring the Rf values.

paper raft technique: A technique created to promote development of single cells, which are taken from suspension cultures, in which cells are placed onto filter-paper squares set on actively growing callus (nurse tissue). Growth factors and nutrients from the callus tissue diffuse through the filter paper, promoting cell growth and development. (Muir, 1973)

paracentric inversion: A rotation of a segment of chromosome a full 180°, with the centromere beyond inversion, which happens all within one arm of the chromosome.

paragenetic: Changes within the eukaryotic chromosomes that effect the expression but not the constitution of the genes concerned. For example, V-type position effects, unstable loci in plants. (Brink, 1962)

parallel mutations: Closely similar mutations occurring in two or more species of the same genus and affecting homologous genes and homologous processes. Sometimes mutations in nonhomologous genes are included under parallel mutations.

paralogous genes: Homologous genes that have arisen through a gene duplication and that have evolved in parallel within the same organism. *Compare* ORTHOLOGOUS GENES.

paramutation: A mutation in which one allele in a heterozygous condition changes its partner allele permanently. (Brink, 1958)

paranemic joint: A region in which two complementary sequences of DNA are associated side by side instead of being intertwined in a double helical structure. pBR322 is one of the standard plasmid helical structures.

paraphyletic: A population consisting of a species and some but not all of its descendants.

pararetrovirus: A viral retroelement whose encapsidated genome is made of DNA.

parasexual reproduction: Any reproductive process that generates recombinants without the formation of gametes. (Pontecarvo, 1949)

parasitic DNA: *See* SELFISH DNA.

parental imprinting: The differential activation and deactivation of maternal and paternal genes.

parentals: Nonrecombinants.

partial denaturation: The partial unwinding of the double helix. Those regions that remain intact late are probably GC-rich, since G-C base pairs, held together by three hydrogen bonds, are more stable than A-T base pairs (two hydrogen bonds).

partial (incomplete) digest: The treatment of a DNA sample with a restriction enzyme for a limited period so that only a proportion of the target sites in any individual molecule are cleaved. Partial digests are often performed with four base-pair cutters to provide an overlapping collection of DNA fragments for use in the construction of a gene bank.

partial diploids: Bacterium containing two alleles for a particular locus owing to gene transfer by transduction or sexduction.

partial dominance: A case where dominance is not complete, and the hybrid is intermediate in type between the two parents. An F_2 1:2:1 ratio is expected in a monohybrid. *Compare* INCOMPLETE DOMINANCE.

particulate inheritance: The model proposing that genetic information is transmitted from one generation to the next in discrete units (particles), so that the character of the offspring is not a smooth blend of essences from the parents. *Compare* BLENDING INHERITANCE. *See* MENDEL'S LAWS.

particulate radiation: Radiation that emits particles when passed through a medium.

particulate theory: A conclusion drawn from Mendel's work about the particulate nature of the gene.

particulate variation: *See* DISCONTINUOUS VARIATION.

passage: The transfer of a sample of cells from one culture vessel to another after growth. It is synonymous with the term *subculture*.

passage number: The number of subcultures performed after the original isolation of the cells from a primary source.

passenger DNA: Foreign DNA incorporated into a plasmid.

paternal chromosomes: Chromosomes derived from the father.

paternal inheritance: Inheritance of the cytoplasmic organelles from a male parent to the offspring. First discovered by Russel (1980) in an angiosperm *Plumbago zeylamica*.

path diagram: A modified pedigree showing only the direct line of descent from common ancestors.

patriclinous inheritance: Inheritance in which all offspring have the nucleus-based phenotype of the father.

pBR322: *E. coli* plasmid that is used very frequently as a cloning vector for the propagation of other gene sequences. It grows in the cell under relaxed control and thus, can be harvested in large numbers. It contains genes for tetracycline resistance and ampicillin resistance.

pedigree: A table, chart, or diagram representing the ancestry of an individual.

pedigree method of breeding: A system of individual plant selection during the segregation generations of a cross, where the progeny plants usually are separately spaced, and the pedigree of a particular selection is known.

pedigree selection: Selection procedure in a segregating population in which progenies of selected F_2 plants are reselected in succeeding generations until genetic purity is reached.

Pelham box: A conserved, short DNA sequence (14 bp) in the 5′ noncoding region, immediately upstream from the TATA box, that serves as the promoter for transcription of heat shock genes.

penetrance: The extent to which a phenotype is expressed, measured as the proportion of individuals with a particular genotype that actually display the associated phenotype. (Vogt, 1926)

penetrance, reduced: A reduction in the number of individuals in any one phenotype, though the number in the corresponding genotype is not reduced; i.e., the particular gene or genes involved, though present, are not always manifest.

pentaploid: Having five sets (genomes) of chromosomes; chromosome number of 5x.

peptide nucleic acid (PNA): It has the ability to replicate itself and catalyze chemical reactions as RNA does. It is a very simple molecule consisting of a peptide (polyamide) backbone composed of H-(2-amino ethyl) glycine units to which nucleobases are attached by carbonyl methylene linkers. It can serve as a template both for its own replication and for the formation of RNA from its subcomponents (Eghlom et al., 1993; Wittung et al., 1994). Bohlar, Nielson, and Orgel (1995) have found that a polycyticidine decamer of PNA can act as a template for oligomerization of activated guanosine mononucleotides.

peptidyl transferase: The enzyme activity responsible for peptide-bond synthesis during translation.

peptidyl-site: The site on the ribosome to which the tRNA attached to the growing polypeptide is bound during translation. Also known as *P-site. Compare* AMINOACYL-SITE (A-SITE).

pericentric inversion: Refers to an inversion that does include the centromere, hence involves both arms of a chromosome.

periclinal chimera: One tissue completely surrounding another type of tissue.

permanent cell line: A cell line that survives in culture through an unlimited number of generations.

permanent heterosis: A hybrid condition maintained by a balanced lethal system; e.g., *Oenothera lamarckiana* is a permanent hybrid of *gaudensvelans* complex.

permissive conditions: The conditions that allow conditional lethal mutants to survive.

permissive euchromatin: Describes most euchromatin, which is transcriptionally silent but potentially active. It is immediately responsive to specific signals such as hormones. It includes luxury genes.

permissive host: A host cell in which a particular strain of virus can reproduce to yield progeny virus. Cells that do not support such growth are defined as nonpermissive.

permissive strain: The strain of cell capable of acting as a permissive host.

pest: A disease-inducing organism, often applied to an organism that causes disease or attacks agricultural crops.

Ph locus: In the presence of the *Ph* gene, chromosome pairing in wheat is restricted to pairing between homologous chromosomes; no pairing occurs between homoeologous chromosomes.

phage: *See* BACTERIOPHAGE.

phage induction: The transition from the prophage state for a lytic growth of a temperature phage in response to an external stimulus. Induction results in the initiation of transcription of phage genes, excision of the prophage from the host chromosome, and the synthesis of phage genomes and capsid proteins.

phase, frame: Nucleotide triplets are "in phase" with one another if they can be read in groups of three from an AUG initiation codon and can, therefore, act as codons for amino acids during translation.

phase-contrast microscope: A light microscope that uses the differences in the refractive index between organelles and their surrounding cytoplasm to obtain contrast. Thus, organelles can be seen without staining, and their size, shape, and motion can be observed in the living state.

phasmid: A plasmid gene-cloning vector carrying the attachment site (*att* P) from lambda phage. It is thus able to insert into a phage genome just as lambda phage inserts into the bacterial genome. (Kahn and Helinski, 1978)

phene: Phenotypic character controlled by genes. Note, however, that in fact all "genetically determined" characteristics are more accurately understood as resulting from an interaction of gene expression and environmental influence. *See* PHENOTYPE. (Johannsen, 1909)

phenocopy: A nongenetic, environmentally induced imitation of the effects of a specific genotype. (Goldschmidt, 1915)

phenodeviants: Phenotypes that deviate from the population (or species) mean (or norm), owing to special gene combinations, for instance extreme homozygosity. (Lerner, 1954)

phenogenetics: *See* DEVELOPMENTAL GENETICS. (Fischer, 1939)

phenogroup: The antigenic determinants that are transmitted as a block from parents to offspring.

phenological isolation: Biological isolation of one species or group from the other members of the same genus by reason of differences in time of flowering (early or late).

phenotype: The observable characteristics displayed by a cell or an organism. Phenotype changes continuously during the developmental processes of an individual. Phenotype is a result of a specific interaction between a genotype and an environment. (Johannsen, 1903)

phenotypic flexibility: The ability of a genotype to vary its expression in different environments. This is a character of considerable adaptive significance in most plants. Species in a genus, as also populations within a species, show differences in the phenotypic flexibility, indicating that this attribute is genetic based and subjected to the control of the usual processes of natural selection. (Thoday, 1953)

phenotypic ratio: The ratio of different phenotypes in a segregating generation. The phenotypic ratio in F_2 of a dihybrid cross is 3 tall : 1 dwarf.

phenotypic sex determination: Sex determination by nongenetic means.

phenotypic value: The numerical description of a phenotype with respect to a particular quantitative character, measured in metric units.

phenotypic variation: When individuals with the same genotype develop in different environments, their phenotypes may be quite different.

ϕX174: A bacteriophage that attacks *E. coli*. The nucleic acid within the viral particle is a single-stranded DNA circle of 5,375 bases. The replicative form is a double-stranded circle and was the first complete DNA molecule to be sequenced. This sequence revealed the presence of overlapping genes.

photoreactivation: A light-induced reversal of ultraviolet light causing injury to cells. (Kelner, 1949)

phyletic evolution: Change that occurs sequentially in a single line of ascent. *Phyletic gradualism* is a slow, directional change of the gene pool producing a transformation of the entire population from one species to another through a series of graded, intermediate forms. *Phyletic speciation* is the process of speciation caused by the gradual change in the genetic constitution of a population, without the population splitting into demes and without any increase in the number of species produced by that population at any one time. *Phyletic succession* is the gradual replacement of one species by another without branching. (Simpson, 1944)

phylogenetic tree: A diagram indicating the supposed sequence of historical relationships linking together species or high taxa.

phylogeny: The evolutionary development of a species or other taxonomic group. (Haeckel, 1866)

physical containment: The use of physical barriers to prevent the escape of genetically engineered organisms from the laboratory into the environment. Precautions employed include the maintenance of negative air pressure within the laboratory and the use of fume hoods with HEPA filters on their exhausts.

physical map: Mapped elements are assigned to physical locations ranging from a whole chromosome to the entire DNA sequence.

physical mutagen: A mutagen of nonbiological or nonchemical origin such as electromagnetic radiation (UV light, X-rays, and γ-rays), particle radiation (electrons, neutrons, protons, ∝-particles, and ß-particles), heat, or mechanical force.

physiological race: Pathogens of the same species and variety, which are structurally similar but which differ in physiological and pathological characteristics, especially in their ability to parasitize varieties of a particular host.

physiological resistance: A type of resistance due to physiological or protoplasmical characters.

physiological suppression: A defect in one chemical pathway due to a mutation is circumvented by another mutation in another metabolic pathway. Thus, when a mutation shuts one pathway, another mutation opens a different pathway leading to the same result.

planosome: A supernumerary chromosome resulting from nondisjunction of a bivalent in meiosis.

plant breeding: The application of genetic analysis to development of plant lines better suited for human purposes.

plant cloning vehicle: It is a vector used in gene manipulation to carry foreign DNA and insert it in an inheritable manner into plant cells, possible vectors include the Ti plasmid of *Agrobacterium tumefaciens* and the DNA plant virus.

plant DNA virus: A virus that contains DNA and infects plants. There are two groups: *Caulimnoviruses* (with dsDNA) and *Gerinivirus* (with ssDNA).

plant genetics: A branch of genetics that deals with the inheritance of characters in plant species.

plant introduction: (1) The transport of a collection of seeds, plants, or vegetative propagating materials from one ecological area into another; (2) a collection of seeds, plants, or vegetative propagating materials that have been transported from one area into another.

plant tissue culture: The growth or maintenance of plant cells, tissues, organs, or whole plants *in vitro*.

plant-animal mutualism: This occurs when one plant and one animal species develop a considerable degree of dependence on each other such that elimination of one species leads to a marked numerical response in the other species.

planting efficiency: This is a term which originally encompassed the terms, "attachment efficiency," "seeding efficiency," "cloning efficiency," and "colony-forming efficiency," and which is now better described by using one or more of them in its place as the term "planting" is not sufficiently descriptive of what is taking place. *See* ATTACHMENT EFFICIENCY; SEEDING EFFICIENCY; CLONING EFFICIENCY; COLONY-FORMING EFFICIENCY.

plasmagene: A self-replicating, cytoplasmically located gene. (Winkler, 1920; Darlington, 1939b)

plasmid: A usually circular 2 to 500 kb long piece of DNA, primarily a replicon that can exist independent of the host chromosome and often found in bacterial and some other types of cells. It is stably inherited in extracellular states. Most, but not all, plasmids are dispensable. Some plasmids have the capacity to integrate themselves into the host chromosomes (episomes). (Lederberg, 1952)

plasmid amplification: A method, involving incubation with an inhibitor of protein synthesis, aimed at increasing the copy number of certain types of plasmid in a bacterial culture.

plasmid incompatibility: The ability of two different plasmids to coexist in the same cell in absence of selection pressure. Twenty-five incompatibility groups of plasmids have been defined in *E. coli.* (Scaife and Gross, 1962; Echols, 1963)

plasmid purification: The separation of plasmid DNA from other forms of prokaryotic DNA and cellular components. The first stage is cell lysis.

plasmid replicon: A base sequence that acts as an "origin" of replication that has been isolated from a plasmid and used in the construction of cloning vehicles.

plasmid rescue: A technique used for transforming *Bacillus subtillis* in which an inactive donor plasmid interacts with a resident plasmid to effect a transformation. In practice, foreign DNA is ligated to monomeric vector DNA, and the recombinant inserted into cells containing a homologous plasmid. The resident plasmid results in transformation. (Contente and Dubnau, 1979)

plastid inheritance: Inheritance that is governed by chloroplast genes.

plating: A technique used to obtain pure cultures of microorganisms (bacteria, yeasts, and fungi) that produce a distinct colony when grown on a solid medium (a nutrient medium solidified with agar or a similar agent in a petri dish).

playback experiment: An experiment, which describes the retrieval of DNA which has a hybrid with RNA, to check that it is nonrepetitive by a further reassociation reaction.

plectonemic winding: The pairing and intertwining of the two strands in the classical double helix of DNA.

pleiotropic: A gene having multiple effects, usually in parts of the organism not obviously related (e.g., flower color and seed coat color). (Plate, 1910)

pleiotropic gene: A gene that affects more than one (apparently unrelated) characteristic of the phenotype.

pleiotropic mutation: A mutation that has effects on several different characters.

pleiotropy: The ability of a single gene to produce a complex phenotype that consists of two or more distinct characteristics.

ploidy: A variation in the number of chromosome sets per cell.

plus and minus strands: Conversion for designating strands of DNA or RNA and messenger strands of DNA or RNA. Messenger RNA and other strands with the same sequence polarity are called plus strands, whereas the coding strand is a minus strand. Some single-stranded viruses are genomes of, for example, plus strand DNA, which must be first replicated to provide minus strand DNA from which transcription can occur.

pMB9: A small multicopy plasmid that carries a tetracycline resistance gene.

PMC: *See* POLLEN MOTHER CELL.

point mutation: A mutation that results from a single nucleotide alteration in a DNA molecule. (Bridges, 1923)

Poisson distribution: A statistical equation that describes the process of sampling in situations where the number of events per sample is potentially very large but in practice is very small.

Pol: The gene that encodes reverse transcriptase in certain eukaryotic RNA viruses.

Pol I (RNA polymerase I): Enzyme responsible for catalyzing the synthesis of large ribosomal RNA (28S rRNA).

Pol II (RNA polymerase II): Enzyme responsible for catalyzing the synthesis of heterogeneous nuclear RNA (hnRNA) and messenger RNA (mRNA).

Pol III (RNA polymerase III): Enzyme responsible for catalyzing the synthesis of small ribosomal RNA (5S rRNA) and transfer RNA (tRNA).

polar gene conversion: A gradient of conversion frequency along the length of a gene.

polar nuclei: Two centrally located nuclei in the embryo sac that unite with the sperm nucleus of the pollen grain in a triple fusion. In certain seeds, the product of this triple fusion develops into the endosperm.

polarity: The effect of a mutation in one gene in influencing the expression (at transcription or translation) of subsequent genes in the same transcription unit. A mutation in any structural gene of an operon that reduces the wild-type distal to it is known as *polar mutation.* (Franklin and Luria, 1961; Jacob and Monod, 1961b)

pollen culture: The cultivation of pollen grains *in vitro* for producing haploid plants (generally referred to as anther culture).

pollen grain: The young male gametophyte of a flowering plant; it is surrounded by the microspore wall.

pollen mother cell: The microsporocyte (2n) in plants immediately before the reduction division. It is the cell where meiotic chromosomes are readily studied.

pollen parent: The plant supplying the pollen for a hybrid.

pollination: The transfer of pollen from the anther to a stigma.

poly(A) polymerase: The enzyme responsible for polydenylation of a eukaryotic mRNA molecule. Most eukaryotes have two poly(A) polymerases; one in the nucleus, which adds an oligo(A) stretch in the nucleus, and another in the cytoplasm, which extends the tail to its full length.

poly(A) site selection: In eukaryotes, tissue-specific selection (in certain transcription units) of alternative poly(A) sites and production of multiple messenger RNAs from one primary transcript.

poly(A) tail: The initially long sequence of adenine nucleotides at the 3′ end of mRNA added after transcription.

polyacrylamide gel electrophoresis (PAGE): A method for separating nucleic acid or protein molecules according to their molecular size. The molecules migrate through the inert gel matrix under the influence of an electric field. In the case of protein, PAGE detergents, such as sodium dodecyl sulphate, are often added to ensure that all molecules have a uniform charge.

polyadenylation: The posttranscriptional addition of 50 to 250 adenosine residues to the 3′ end of a eukaryotic mRNA molecule. Poly(A) tail is associated with proteins and undergoes size reduction in the cytoplasm in an age-related process (Edmond and Abrams, 1960). The following consensus sequence is located 5 to 20 bp downstream from the poly(A) addition site:

```
        G
    T T   N N N T T T T T T
        A
```

polyallel crossing: The crossing of each inbred line with every other line or with certain other inbred lines; an extension of diallel crossing to cover the simultaneous comparison of more than two sires.

polycentric: A chromosome or chromatid having several centromeres.

polycistronic message: A molecule of messenger RNA that is the product of transcription of a series of tandemly located genes on the DNA. Although some polycistronic messages are later processed by cleavage into separate messages, some are translated intact. The messages of bacterial operons such as the *lac* operon of *E. coli* are polycistronic.

polycistronic mRNA: An RNA molecule, mostly in prokaryotes, that contains information from more than one cistron.

polycloning region: A segment of DNA in a cloning vector, such as a plasmid, that contains many different restriction enzyme cleavage sites.

polycross: Random open-pollination among a set of lines in isolation under a polycross nursery where lines are highly replicated and grown in all directions to allow free pollen-flow. The bulk seed finally collected forms polycross seeds. Progenies from individual lines provide information on the breeding value of each line involved. Polycross is a natural substitute for topcross.

polycross progeny: Progeny from a selection, line, or clone outcrossed to other selections growing in the same isolated polycross nursery.

poly-dA/poly-dT technique: A method of inserting foreign DNA into a vehicle by making 5′ poly-dA and 5′ poly-dT tails on the vehicle and on foreign DNAs. Also feasible with poly-dG and poly-dC.

polyembryony: The development of more than one embryo in a seed, or multiple-embryo development in cultures.

polyethylene glycol (PEG): A fusion-inducing agent (fusogen) for agglutinating protoplasts that is used in somatic hybridization. This compound is also sometimes used in media as a nonmetabolite osmoticum and is available in various molecular weights, ranging from 1000 to 6000. PEG 4000 and PEG 6000 are commonly used to promote cell or protoplast fusion and to facilitate DNA uptake in the transformation of organisms such as yeasts. PEG is also used to concentrate solutions by withdrawing water from them. PEG has the formula $HOCH_2$ $(CH_2$ $OCH_2)$ $\times$ CH_2OH.

polygenes: These are two or more different pairs of alleles with a presumed cumulative effect that governs such quantitative traits as size, pigmentation, and intelligence, among others. Those contributing to the traits are termed contributing (effective) alleles; those appearing not to do so are referred to as noncontributing or non-effective alleles. (Plate, 1913; Mather, 1941)

polygenic balance: The harmonious balance, between the systems of polygenes in a species and the natural environment of that species, produced as a result of natural selection and enabling the organisms to develop normally and competitively.

polygenic inheritance: Inheritance involving alleles at many genetic loci, which interact with environmental factors. Polygenic refers to the many genetic factors as opposed to the environmental component. *See* QUANTITATIVE INHERITANCE; MULTIPLE FACTOR HYPOTHESIS.

polygenic trait: A trait not readily separated into discrete groups but that may be continuously variable and often have a normal distribution curve. Such a trait is controlled by more than one pair of genes and may also be modified by environmental factors.

polyhaploid: A haploid plant produced from a polyploid plant; it contains a multiple of the basic (X) chromosome number for that species. It has more than one genome. (Katayama, 1934)

polyhybrid: *See* MULTIHYBRID. (DeVries, 1901)

polyhybrid cross: A cross that differs in several gene pairs.

polylinker: An oligonucleotide DNA sequence that effectively joins DNA sequences colinearly.

polymerase: An enzyme that catalyzes the formation of DNA or RNA.

polymerase chain reaction (PCR): A technique in which cycles of denaturation, annealing with primer, and extension with DNA polymerase, are used to amplify the number of copies of a target DNA sequence by more than 10^6 times. (Saiki et al., 1985)

polymeric genes: Two dominant alleles that have a similar effect when they are separate but that produce an enhanced effect when they come together, resulting in a 9:6:1 ratio. (Nilsson-Ehle, 1908)

polymerism: A phenomenon in which nonallelomorphic genes having identical and cumulative action interact. The genes involved in such interactions cannot be identified individually. These interactions may be expressed statistically as components of genetic variance.

polymorphic DNA: A variable DNA sequence that can exist in a number of different but related forms.

polymorphism: The simultaneous occurrence of more than one biological property in a species with a frequency that cannot be explained by recurrent mutation. (Ford, 1940)

polyneme hypothesis: The proposition that a chromatid contains more than one duplex of DNA running in parallel. This has been largely superceded by the present understanding that a single duplex of DNA is present in a chromatid, but that it is elaborately packaged by supercoiling and looping-out from a central axis. So, only at a superficial level can separate gene loci be regarded as being in purely linear order along a chromosome length.

polynucleotide: A linear sequence of nucleotides in DNA or RNA.

polynucleotide phosphorylase: An enzyme that can polymerize diphosphate nucleotides without the need for a primer. The function of this enzyme *in vivo* is in its reverse order as an RNA exonuclease.

polypeptide: A compound containing amino acid residues joined by peptide bonds. A protein may consist of one or more specific polypeptide chains.

polypeptide elongation: The process of adding amino acids to a growing polypeptide chain.

polypeptide elongation factors: The proteins that play an ancillary role in the elongation step of an amino acid molecule. EF-Ts, EF-Tu, and EF-G are the elongation factors of prokaryotes, and eEF1α, eEF1ß, eEF1γ, and eEF2 are elongation factors of eukaryotes. (Lucas-Lenard and Lipman, 1966)

polyploids: Organisms with more than two chromosome sets. (Dermen, 1940; Winkler, 1916)

polyprotein: A translation product that consists of a series of proteins joined into a single polypeptide and which is processed by proteolytic cleavage to produce the mature proteins.

polyribosome: An mRNA molecule in the process of being translated simultaneously by several ribosomes. (Warner, Rich, and Hall, 1962)

polysome: *See* POLYRIBOSOME.

polysomic: An individual having either a single extra chromosome or one pair of extra chromosomes in the diploid set.

pooled libraries: For long-term storage, phage particles or bacterial colonies on bacteriological plates are eluted in a mixed pool, which contains representatives of every different recombinant on the plate. Also known as *amplified libraries*.

population genetics: The study of the frequencies of genes and genotypes in a Mendelian population.

population structure: The sum of all the factors that govern the pattern in which gametes from various individuals unite with each other.

position effect: A phenotypic effect dependent on a change in position on the chromosome of a gene or group of genes. It may produce variegation, chimera, or a mosaic phenotype. (Sturtevant, 1925)

position-effect variegation: Variegation caused by the inactivation of a gene, in some cells through its abnormal juxtaposition with heterochromatin.

positive assortative mating: A situation in which like phenotypes mate more commonly than expected by chance.

positive control: Activation of an operator site by a regulatory protein that is produced by a regulator site.

positive interference: This exists when the occurrence of one crossover reduces the probability that a second will occur in the same region. *Compare* NEGATIVE INTERFERENCE.

positive regulator proteins: The proteins required for the activation of a transcription unit.

positive selection: The situation where cells carrying a certain gene can be detected because the activity of that gene is essential for cell growth under certain conditions; e.g., antibiotic resistance genes are positively selectable.

positive supercoiling: Describes the coiling of the double helix in space in the same direction as the winding of the two strands of the double helix itself.

postadaptive mutation: A mutation that occurs as a direct response to an environmental stress and that enables the organism to adapt to that stress. A discredited idea.

postmeiotic segregation: The segregation of two strands of a duplex DNA that bear different information (created by heteroduplex formation during meiosis) when a subsequent replication allows the strands to separate.

postreplicative repair: A DNA repair process initiated when DNA polymerase bypasses a damaged area. Enzymes in the *rec* system are used. In bacteria, it takes place as follows: (a) replication proceeds past the DNA lesion in the parental strand, leaving gaps 500 to 1,000 nucleotides wide in the daughter strand; (b) the gaps in the daughter strand are filled with material from the parental strands by single-strand exchanges between the sister duplexes; (c) the gaps (about 15,000 nucleotides wide) formed in the parental strands as a result of the recombination process are filled by replication repair. In mammalian cells, the gaps (about 100 nucleotides wide) are formed by a bypass mechanism similar to bacteria, but although recombinational exchanges are involved in gap filling, the gaps are less frequent and/or much shorter than those in bacterial DNA. Postreplication repair is a error-prone process. (Rupp and Howard-Flanders, 1968)

posttranscriptional modifications: Changes in eukaryotic mRNA made after transcription has been completed. These changes include additions of caps and tails and removal of introns. *See* RNA PROCESSING.

posttranslational modifications: Changes in polypeptide after translation has been completed. Such modifications may lead to a change in phenotype. For example, posttranslational modifications lead to isozymes. *See* EPIGENETIC MODIFICATIONS.

postzygotic isolating mechanism: Any one of several mechanisms that keeps populations reproductively isolated from each other even though fertilization and hybrid zygotes may form. These hybrids are either sterile, nonviable, or so weak that they do not survive.

potence: The property of a group of polygenes corresponding to the degree of dominance of a major gene.

potency (of a gene): The capacity of a gene for manifesting its presence. *Partial potency* is synonymous with *incomplete dominance*. *Compare* EXPRESSIVITY and PENETRANCE.

potential heterozygosity (H): Having more than twice the normal haploid number of chromosomes.

potentiation: When two or more factors, physical or chemical, together cause an increase in mutation frequency beyond that caused by each factor individually.

preadaptive mutation: A mutation that occurs before a specific environmental stress is encountered. When the stress occurs, the mutation, which confers a selective advantage on the organism, is selected for.

predator-prey adaptations: The coexistence of two species delicately depends upon the genetic composition of the competing populations due to a counterevolutionary equilibrium.

predisposition: The tendency of some nongenetic factors, such as abiotic stresses (water-stress, temperature-stress, etc.), to affect susceptibility of plants to diseases, hence associated with the disease triangle.

preempter stem: A configuration of leader transcript mRNA that does not terminate transcription in the attenuator-controlled amino acid operons.

pre-mRNA: The primary, unprocessed transcript of a protein coding gene. Pre-mRNA molecules are much larger (up to 10^7 daltons) than messenger RNAs (10^5 to 10^6 daltons). *See* HETEROGENEOUS NUCLEAR RNA. (Georgiev and Mantieva, 1962; Scherrer and Darnell, 1962)

prepotency: The capacity of a parent to impress its characteristics on its offspring, such that they are more alike with parent and with each other than usual.

prepriming complex: The structure, consisting of a series of proteins attached to the origin of replication, that initiates DNA replication in *E. coli*.

prepriming protein: A protein that recognizes primer initiation sites.

pre-rRNA: The primary, unprocessed transcript of a gene or group of genes specifying rRNA molecules. In bacteria, no large, molecular-weight precursor containing the joint sequences of 5′ 16S-(tRNA)-23S-5S 3′ accumulates as the growing RNA chain is cleaved by RNase III to yield immediate precursors of the mature rRNA molecules. The precursors are called p16, p23, and p5. Eukaryotic rRNA is transcribed as a long precursor molecule that is converted to mature 18S, 5.8S, and 28S rRNA by a series of endonucleolytic cleavages.

presence and absence hypothesis: The discredited idea that the normal phenotype is due to the presence of a factor and the mutant phenotype to its absence.

preteome: An emerging field of research that concentrates on the study of proteins produced by a cell in order to define various processes in molecular terms; whereas genome study of DNA sequences is producing more data than they can put a function to. (Nowak, 1995)

pre-tRNA: The primary, unprocessed transcript of a gene or group of genes specifying tRNA molecules. Prokaryotic pre-tRNA is composed of about 120 nucleotides and eukaryotic pre-tRNA of 300 to 400 nucleotides.

prezygotic isolating mechanism. Any one of several mechanisms that keep populations reproductively isolated from each other by preventing fertilization and zygote formation.

prezygotic isolation mechanisms: Mechanisms active before the egg has become fertilized; usually applied to isolation mechanisms that interfere with fertilization of the egg.

Pribnow box: A relatively invariable six nucleotide DNA sequence, TATAAT, located in prokaryotic promoters upstream (at position – 10) from the start codon. It has the sequence:

5′	T	A	T	A	A	T	G	3′
3′	A	T	A	T	T	A	C	5′
	–12	–11	–10	–9	–8	–7	–6	

primary cellular function: Synthesis of a transcript—mRNA, tRNA, or rRNA.

primary constriction: A constriction that is determined by and associated with the centromere region.

primary culture: A culture started from cells, tissues, or organs taken directly from organisms. A primary culture may be regarded as such until it is successfully subcultured for the first time. It then becomes a "cell line."

primary nondisjunction: The failure of homologous chromosomes to separate at anaphase I of meiosis. This process leads to production of aneuploids.

primary protein structure: The sequence of polymerized amino acids in a protein.

primary transcript: The immediate product of transcription of a gene or group of genes, which will subsequently be processed to produce the mature transcript(s).

primary trisomic: A trisomic in which the additional chromosome is one of the normal chromosomes of the complement. (Blakeslee, 1921)

primase: The RNA polymerase enzyme that synthesizes the primer needed to initiate replication of a DNA polynucleotide. (Lark, 1972; Rowen and Kornberg, 1978)

primed *in situ* hybridization (PRIN): Involves annealing of unlabeled primers followed by incubation with a reaction mixture that includes labeled nucleotides and DNA polymerase.

primer RNA: A short RNA sequence of about ten nucleotides that is synthesized by a primase on the lagging strand during DNA replication. These short lengths of RNA are then extended by DNA polymerase to yield the *Okazaki fragments,* from which the primer RNA molecules are then erased and the pieces "stitched" together by DNA ligase to yield a continuous, new DNA strand.

primosome: A complex of two proteins, a primase, and a helicase that initiates RNA primers on the lagging DNA strand during DNA replication. Primosomes are an integral part of the replication machinery. (Low, Arai, and Kornberg, 1981)

principle of independent assortment (segregation): *See* MENDEL'S SECOND LAW.

principle of segregation: *See* MENDEL'S FIRST LAW.

probability of ultimate fixation (U_p): The probability that a given allele with frequency p will eventually become fixed within a population.

probe: In recombinant DNA work, a radioactive nucleic acid complementary to a region being searched for in a restriction digestor genome library.

process versus discard decision: This refers to the decision taken by pre-mRNA processing machinery whether to process a particular molecule in one cell type or to discard it in another cell type. There is no evidence for existence of such a mechanism for nuclear RNA.

processed gene: A pseudogene in the eukaryotic genome that lacks introns but contains a poly(A) sequence at its 3′ end. These characteristics suggest that such genes are derived from reversed copying of a processed mRNA molecule, perhaps involving a reverse transcriptase. *See* PSEUDOGENE.

processed pseudogene: A pseudogene whose sequence resembles the mRNA copy of a parent gene and that probably arose by integration into the genome of a reverse transcribed version of the mRNA.

processing: (1) Modification of RNA following initial transcription, including capping, tailing, intron excision, and exon splicing; (2) modification of an antigen by macrophages to a form such that the antigen appears in the macrophage surface membrane and is then stimulatory to the appropriate lymphocytes.

processive enzymes: These enzymes continue to act on a particular substrate; that is, they do not dissociate between repetitions of the catalytic event.

processivity of an enzyme: The ability of an enzyme to repetitively continue its catalytic function without dissociating from its substrate.

processivity of DNA replication: The frequency with which an enzyme dissociates from template-primer after addition of a nucleotide to the growing chain end.

producer gene: A eukaryotic structural gene that produces a pre-mRNA molecule, which after going through some processing steps, becomes mRNA. A producer gene may be under the control of several receptor genes.

progeny selection: Selection based on progeny performance.

progeny testing: The determination of breeding value of a parent based on its progeny-performance.

prokaryote: An organism or cell that lacks a nucleus bounded by a membrane, as well as specialized membrane-bound organelles. *Compare* EUKARYOTE. (Chatton, 1925)

promiscuous DNA: The occurrence of some DNA sequences in more than one cellular compartment. It suggests that DNA may have been exchanged between organelles, or between organelles and the nucleus. (Ellis, 1982)

promiscuous plasmid: A plasmid that is capable of promoting its own transfer to a wide range of Gram-negative bacteria and being stably maintained in these diverse hosts. These plasmids generally belong to the incompatibility classes known as P and Q.

promoter: The nucleotide sequence, upstream of a gene, to which RNA polymerase binds in order to initiate transcription. Although the binding sites for RNA polymerase generally extend from about 45 bp upstream from the startpoint of transcription to about 20 bp downstream, promoter recognition in bacteria is governed by two regions—the recognition region (-35 region) and the Pribnow box (-10 region). In eukaryotes, several DNA sequences upstream from the initiation site are important for recognition and positioning of RNA polymerase II transcribing mRNA. Polymerase III recognition and positioning involves internal promoters to the 5S and tRNA genes. The controlling elements for transcription by polymerase I (rRNA genes) are not yet identified. (Jacob, Brenner, and Cuzin, 1964)

proofreading: The ability of a DNA polymerase to correct misincorporated nucleotides as a result of its $3'$ to $5'$ exonuclease activity.

prophage: The integrated form of the DNA molecule of a lysogenic phage.

prophase: The first stage of nuclear division including all events up to, but not including, the arrival of the chromosomes at the equator of the spindle. (Strasburger, 1884)

prophase I: The longest period of meiosis I, during which chromosomes condense and homologous chromosomes pair, undergo crossing-over, and then separate-out but still close at the ends. It includes substages leptotene, zygotene, pachytene, diplotene, and diakinesis.

prophase II: A very brief period that may exist between meiosis I and meiosis II. This stage is omitted in organisms that omit interkinesis (a short resting stage between meiosis I and meiosis II).

protamine: A low molecular-weight protein, lacking a prosthetic group but rich in arginine content, found in association with nuclear DNA in certain cells.

protandry: The differential maturation period of reproductive organs where anthers mature earlier than pistils, reverse of *protogyny.*

protease: An enzyme that degrades protein.

protein biosynthesis: The production of protein from its constituent amino acids using mRNA as a template. This takes place in association with specific organelles known as ribosomes.

protein fingerprinting: A technique that produces a pattern of protein fragments (usually resolved on a two-dimensional electrophoretic gel) generated by cleavage with an enzyme such as trypsin. A protein, having a repeated motif of amino acids with characteristic spacing of cysteines that may be involved in binding zinc, is characteristic of some proteins that bind DNA and/or RNA. It produces zinc protein fingerprints.

protein sequence phylogenetics: The technique used to determine the genetic relationship between various organisms. It is based on the determination of the amino acid sequence of a comparable protein, often cytochrome C.

protein targeting: A step in importing proteins from their site of synthesis to their site of function in which binding of proteins occurs with specific organelle membranes.

proteinoids: Various amino acids in certain conditions of temperature and pH are referred to as proteinoids or *thermal proteinoids*, which are thought to be involved in the origin of life.

protoclonal variation: Variation among the plants regenerated from protoplasts.

protogyny: The differential maturation period of reproductive organs where pistils mature earlier than anthers, reverse of *protandry.*

protomer: One of the individual polypeptide subunits that combine to make the protein coat of a virus.

protoplast: A cell from which the cell wall has been completely removed. The term includes plant, bacterial, or fungal cells. *Compare* SPHEROPLAST.

protoplast fusion: A technique for producing hybrids between two cells that would not normally mate. The two cells may belong to the same or different species. The cell walls are removed from the two parent cells to create protoplasts and then fusion of the two cell membranes is promoted, usually by the addition of PEG (polyethylene glycol) and Ca^{2+} ions. These fusogenic agents cause proteins to migrate from certain regions of the two cell membranes and allow areas of naked phospholipid to fuse. Subsequent regeneration of the cell wall allows the propagation of the hybrid organism. If nuclear fusion does not follow cell fusion, then heterokaryons, rather than diploid organisms, are produced. In some crosses, the genetic contribution of the two parents to the stable hybrid can be markedly unequal.

prototroph: An organism such as a bacterium that will grow on a minimal medium. (Ryan and Lederberg, 1946)

provirus: A viral chromosome that has integrated into a host genome. (Temin, 1971)

proximal protein: With reference to the assembly of complex structures, e.g., ribosomes, these are proteins that bind early in the assembly process and whose presence in the partially assembled structure is required for the binding of distal proteins. *See* DISTAL PROTEIN.

pSC101: A small, nonconjugative plasmid that encodes tetracycline resistance. Its origin is not clear as it is thought to have come from a contaminant in a transformation experiment. It was one of the first vectors used in genetic engineering.

pseudoalleles: Nonalleles so closely linked that they are often inherited as one gene but shown to be separable by crossover studies. Pseudoalleles are found to be allelic in complementation tests but nonallelic in recombination tests. Now this term is discarded. (Morgan, Bridges, and Sturtevant, 1928; Lewis, 1948)

pseudoalleles *cis*-arrangement. The arrangement of two mutant alleles in one chromosome strand, with the wild-type alleles in the other.

pseudoalleles *trans*-arrangement: An arrangement with a mutant allele and a wild-type allele in each of two chromosome strands.

pseudoautosomal inheritance: A pattern of inheritance shown by genes that are located on the pairing region of the X and Y chromosomes in individuals having XY mechanism of sex determination.

pseudodominance: The expression (apparent dominance) of a recessive gene at a locus opposite a deletion.

pseudogene: A nucleotide sequence that has similarity to a functional gene but within which the biological information has become scrambled, so that the pseudogene is not itself functional. (Jacq, Miller, and Brownlee, 1977)

pseudoheterosis: Gigantism or lucurience resulting from distant hybridization (interspecific or interracial crossing), often accompanied by a poor or no-seed setting.

pseudoisochromosome: A translocation chromosome with most of one arm homologous to the other arm. It is derived due to exchange of chromosome segments from the opposite arms of the homologous chromosomes. (Caldecott and Smith, 1952)

pseudo-overdominance: *See* TRUE OVERDOMINANCE.

pseudo-self-compatibility: A partial-seed setting following self-pollination in an otherwise self-incompatible plant.

Pst I: A type II restriction enzyme from the bacterium *Providencia stuartii* 164 that recognizes the DNA sequence:

```
                     ↓
5′ C  T  G  C  A  G 3′
3′ G  A  C  G  T  C 5′
       ↑
```

pulse field gel electrophoresis (PFGE): The artificial production of somatic hybrids by the fusion of protoplasts from different plants. A method for fractionating very large DNA fragments produced by restriction endonuclease digestion, involving the migration of fragments through a matrix under the influence of an electric field applied in pulses from different angles.

pulse-chase experiment: These experiments are performed by incubating cells very briefly with a radioactively labeled precursor (of some pathway or macromolecule); then the fate of the label is followed during a subsequent incubating with a nonlabeled precursor.

punctuation codon: A codon that designates either the beginning or the end of a gene.

punctuation equilibrium model: A model of biological evolution according to which species are relatively stable and long-lived, and new species are formed during concentrated outbursts of speciation, followed by differential success of certain species. In other words, the evolutionary pattern is characterized by long periods of stasis of individual species punctuated by outbursts of rapid change.

Punnett square: A tabular treatment used to predict the genotypes of the progeny resulting from a genetic cross in which a number of alleles are followed.

pure breeding line or strain: A population that breeds true for a particular character or set of characters (tall→tall; dwarf→dwarf). (Johannsen, 1903)

purine (Pu): One of the two (adenine and guanine) types of nitrogenous base compounds that are components of nucleotides. Purine base has a two-ring structure.

purity: With reference to sugar beets, the ratio of sucrose to total solids dissolved in sugar beet juice.

puromycin: An antibiotic that inhibits polypeptide synthesis by competing with aminoacyl tRNAs for ribosomal binding site "A."

pyramiding of genes: It is the cataloging of a group of genes (e.g., resistance genes) by marker-assisted selection.

pyrimidine (Py): One of the three (thymine, cytosine, and uracil) types of nitrogenous base compounds that are components of nucleotides. Pyrimidine base has a one-ring structure.

pyrimidine dimer: The compound that results when UV radiation induces the covalent joining of two thymine residues, two cytosine residues, or a thymine and a cytosine.

q: The relative frequency of a gene; when used together with p, it indicates the relative frequency of the recessive alleles.

Q banding: A technique where chromosomes are stained with quinacrin mustard, which gives florescence under UV light and produces characteristic light and dark bands, called Q bands, on chromosomes.

Qb: RNA bacteriophage, best known as the source of an RNA replicase.

Q-beta: A bacteriophage that infects *E. coli.* It contains a single-stranded circular RNA genome. The viral RNA is termed the + strand because it can function as an mRNA molecule and direct synthesis of viral proteins. The + strand also acts as a template for the synthesis of a complementary − strand. The resulting double-stranded molecule, together with the Q-beta replicase, directs the synthesis of a large number of + strands. This autocatalytic replication can be harnessed to produce large quantities of a desired RNA molecule using recombinant RNA technology.

quadriallel analysis: Analysis based on double-crosses obtained by crossing a number of homozygous lines in a diallel fashion, and then crossing the F_1's, so generated, according to a diallel scheme, with the restriction that in any double-cross a homozygous line must not occur more than once as a parent.

quadripartite structure: A structure having four corresponding parts.

quadrivalent: *See* UNIVALENT.

quadruplex: *See* NULLIPLEX. (Blakeslee, Belling, and Farnham, 1923)

qualitative inheritance: Mendelian inheritance, where discrete phenotypes give expected ratios. *See* OLIGOGENIC INHERITANCE.

qualitative trait (phenotype): A trait for which there are a relatively fewer number of discrete phenotypes that can be distinguished by visual observation.

qualitative variation: Discrete variation where phenotypes can be divided into a fewer number of classes.

quantification of genetic variation: The measurement of amounts of genetic variation present in a population. It depends upon resolving power of the techniques used, intensity of efforts made, and diversity of the strains examined.

quantitative genetics: A branch of genetics that deals with the inheritance of quantitative characters. Also called *biometrical genetics* or *statistical genetics*.

quantitative inheritance: The mechanism of genetic control of traits showing continuous variation.

quantitative risk assessment: A methodology for providing estimates of the risk of mutation to an agent.

quantitative trait: A trait, such as height, that has a continuous distribution pattern within a population and that is typically determined by the combined effects of a number of genes. There is a range of phenotypes differing in degree.

quantitative variation: *See* CONTINUOUS VARIATION.

quantum evolution: Rapid evolutionary change, often assumed to be involved in the production of new, higher categories. (Simpson, 1944)

quantum speciation: The budding-off of a new and very different daughter species from a semi-isolated peripheral population of the ancestral species in a cross-fertilizing organism.

quarantine: The control of plant import and export for purposes of disease and pest control and to prevent spread of disease and pests. Control may involve isolation of plant material, sometimes for extended periods of time, to determine its disease status.

quarantine block: An isolated block in which all plants imported into the area are first grown to observe any disease or pest that may be carried by the plants in question and to prevent the spread of disease or pest to plants grown in the locality.

quartet: The four cells (before separation) arising from a spore mother cell. In higher plants they become pollen grains when mature.

quasispecies: (1) A set of related but different individuals in nucleic acid clones during early evolution of life on earth; (2) a heterogeneous mixture of related genomes, as observed in most RNA viruses, that do not consist of a single genome species of defined sequence. (Eigen, 1971)

quaternary hybrid: A hybrid derived from four grandparental types, all of which are different.

quiescent culture: A culture of cells in which active cell growth and replication are minimal but in which cells continue to survive adequately. Confluent cell cultures may behave in this way.

r: (1) That portion of the R plasmid carrying genes for all drug resistance, with the probable exception of tetracycline resistance. *See* R FACTORS and RESISTANCE TRANSFER FACTOR (RTF); (2) in statistics, correlation coefficient.

R factors: Bacterial plasmids carrying genes for drug resistance. *See* r and RESISTANCE TRANSFER FACTOR (RTF.) Also known as *R plasmids*.

R genes: Disease-resistance genes that encode proteins to resist pathogens. Molecular characterization cloning of several R genes has been performed in order to understand molecular events that specify the expression of resistance. (Lamb, 1994)

R loop: The structure formed when an RNA strand hybridizes with its complementary strand in a DNA duplex, thereby displacing the original strand of DNA in the form of a loop extending over the region of hybridizing.

R plasmid: A plasmid containing one or several transposons that bear resistance genes.

r selection: Selection occurring in an unpredictable environment, where physically, density-dependent factors are important; traits for selection include small size, early maturation, single reproduction, large number of offspring, and little parental care.

R_1, R_2, R_3, etc: The first, second, third, etc., generations following any type of irradiation used to induce muttions.

RAC: Recombinant DNA Advisory Committee. A committee of the United States NIH that formulates guidelines governing the performance of genetic engineering experiments.

race: A genetically distinct, and usually geographically distinct, interbreeding division of a species.

race-nonspecific resistance: *See* HORIZONTAL RESISTANCE.

race-specific resistance: *See* VERTICAL RESISTANCE.

radial loop chromosome model: A model for DNA packing in eukaryotic chromosomes according to which a fiber radiates outward from the axis of each chromatid. (Paulson and Laemmli, 1977)

radiation genetics: A branch of genetics that deals with the effects of various types of radiations on chromosomes and genes.

radiation hybrid (RH) mapping: A method of gene mapping that depends upon breakage of chromosomes to determine the distance between DNA markers as well as their order on the chromosomes.

radical substitutions: *See* DRASTIC AMINO ACID SUBSTITUTIONS.

RAM mutants: Ribosomal ambiguity mutants that allow incorrect tRNAs to be incorporated into the translation process.

random amplified microsatellite polymorphisms (RAMPS): A variant of *MP-PCR (microsatellite-primed PCR*) technique that combines a 5′ anchored microsatellite primer with an arbitrary primer, resulting in so-called RAMPS. RAMP technique has been exploited for genetic mapping.

random chromatid segregation: This exists if the crossing-over frequency between the gene centromere in a multivalent (tri- or tetrasome) is such that two chromatids behave independent of each other.

random fixation: The complete loss of one of two alleles in a population, the other allele reaching a frequency of 100 percent under certain calculable circumstances. (Wright, 1931)

random genetic drift: Changes in allelic frequency due to sampling error.

random mating: This occurs when an individual of one sex has the equal probability of mating with any individual of the opposite sex.

random strand analysis: Mapping studies in organisms that do not keep together all the products of meiosis.

randomized block design: An experimental design that controls fertility variation in one direction only.

randomly amplified microsatellite polymorphisms (RAMPO): Blotting and hybridization of RAPDs to microsatellite-complementary probes results in unexpected polymorphic banding patterns. This new and so far untapped, marker source is also known as *randomly amplified hybridizing microsatellites (RAHM)/ randomly amplified microsatellites (RAMS)*.

RAPD markers: Random amplified polymorphic DNAs that serve as markers to detect the presence or absence of primer-binding sites separated by defined lengths of intervening DNA.

RAPDs: Random amplified polymorphic DNAs.

rapidly reassociated DNA: The fraction of DNA that anneals or renatures rapidly after denaturation. The annealing involves the recovery of double-strandedness by specific hydrogen bonding, and its rate in solution is a function partly of the frequency of a particular sequence in the mixture and partly of the nature of the sequences involved. The DNA that reassociates most rapidly is the palindromic sequence material, which can hybridize with itself by the doubling-over of a single-stranded molecule. It is thus called snap-back or fold-back DNA. A fraction that reassociates somewhat more slowly than fold-back DNA, but still much faster than bulk eukaryotic DNA, is highly repetitive and largely composed of centromeric DNA. DNA of intermediate speed of reassociation is termed moderately repetitive DNA and comprises material such as ribosomal and transfer RNA genes, as well as genes coding for histones, which are present in multiple copies. *See* ANNEALING OF DNA.

rare base pairing: Pairing between two purines or pyrimidines in DNA.

rare cutter: (1) Restriction enzymes that recognize eight nucleotide, or longer, segments and generate DNA fragments ranging in size from

50 kb to over 9 mb—for example, Not I, Rsr II, Sfi I, Sty I. Rare cutters are invaluable for long-range physical mapping of genomic DNA; (2) a restriction enzyme which recognizes sites that are infrequently represented.

rate gene: A gene controlling the rate of a certain developmental process. (Goldschmidt, 1917)

rDNA: *See* RIBOSOMAL DNA.

reading frame: A particular sequence of nucleotide triplets (codons) employed in translation. The reading frame depends on the location of the initiation codon; addition or deletion of one or two bases will alter the reading frame, whereas three base deletions or additions will have no effect on the reading frame. *See* READING FRAME SHIFT.

reading frame shift: A mutation that induces the addition or deletion of a base within a coding sequence, resulting in misreading of the code during translation. This is because a DNA sequence is read in triplets (codons) from a fixed point of initiation, and any addition or deletion will lead to all triplets downstream of the mutation differing from the original.

reading mistake: The incorrect placement of an amino acid residue in a polypeptide chain during protein synthesis.

readthrough: In the presence of a suppressor-tRNA, an amino acid can be inserted into a growing polypeptide chain in response to a stop codon. Termination is thus prevented and a longer than usual polypeptide synthesized. This process is known as readthrough, and the protein product is a readthrough protein.

realized heritability: Heritability measured by a response to selection.

reannealing: The spontaneous realignment of two single DNA strands to re-form a DNA double helix that had been denatured.

rearrangement: A structural change of the chromosome.

reassociation kinetics: *See* RENATURATION KINETICS.

reassociation of DNA: The pairing of complementary single strands to form a double helix.

receptor element: A controlling element that can insert into a gene (making it a mutant) and can also exit (thus making the mutation unstable); both of these functions are nonautonomous, being under the influence of the regulator element.

receptor gene: A component of eukaryotic gene regulation with which a specific activator RNA molecule complexes. This interaction activates the receptor gene.

recessive: The allele whose phenotype is not expressed in a heterozygote. A recessive allele expresses its effect in homozygous or hemizygous states. The converse of dominance—a completely recessive allele is expressed only in the homozygous state. Also known as *recessiveness*. (Mendel, 1866)

recessive epistasis: A condition where a homozygous, recessive gene pair masks the effect of another gene.

recessive lethal: An allele that is lethal when the cell is homozygous for it.

recessive phenotype: The phenotype of a homozygote for the recessive allele; the parental phenotype that is not expressed in a heterozygote.

reciprocal altruism: An apparently altruistic behavior done with the understanding that the recipient will reciprocate at some future date.

reciprocal chiasmata: Chiasmata formed between two chromatids in such a way that a double crossing-over occurs in which the other two chromatids are not involved.

reciprocal cross: A second cross of the same individuals in which the sexes of the parental generation are reversed (tall × dwarf and dwarf × tall is a set of reciprocal crosses).

reciprocal recombination: The production of new genotypes with the reverse arrangements of alleles according to maternal and paternal origin.

reciprocal recurrent cross: A recurrent selection breeding system in which genetically different groups are maintained, and in each selection cycle, individuals are mated from the different groups to test for combining ability.

reciprocal translocation: The exchange of segments between two nonhomologous chromosomes.

recircularization: The ligation of the two ends of a linear DNA molecule to re-form a circular plasmid.

recoding: It is a programmed genetic decoding. A minority of mRNAs in bacteria carry instructions that specify an alteration in how the genetic code is to be read-out, so that the meaning of code words is altered. The phenomenon is called recoding and the instructions in the mRNA are known as *recoding signals.* (Gesterland, Weiss, and Atkins, 1992)

recognition site: A sequence of bases within the promoter region that serve to recognize the RNA polymerase molecule. *See* PROMOTER and INITIATION SITE.

recombinant: A cell, derived from a genetic cross, that displays neither of the parental combinations for the alleles under study.

recombinant DNA: A DNA molecule created in the test tube by ligating together pieces of DNA that are not normally contiguous.

recombinant DNA technology: All the techniques involved in the construction, study, and use of recombinant DNA molecules. The recombination process consists of several steps: (a) isolation (or synthesis) of DNA fragments to be cloned; (b) ligation to a cloning vector *in vitro* to produce recombinant DNA molecules; (c) introduction of recombinant DNA into the host for replication (using the replication origin of the vector); and (d) selection of recipient cells that have acquired the recombinant DNA.

recombinant joint: The point at which two recombining molecules of duplex DNA are connected (the edge of the heteroduplex region).

recombinant protein: A polypeptide that is synthesized by the use of a recombinant DNA molecule as template.

recombinant RNA: A term used to describe RNA molecules joined *in vitro* by T_4 RNA ligase. The recombinant RNA molecule can be autocatalytically replicated by Q-beta replicase to produce large amounts of an RNA sequence of interest.

recombinant RNA technology: A technique that involves the construction of recombinant RNA and allows the production of large quantities of RNA by conferring on any RNA molecule the ability to be replicated in test tube reactions. (Miele, Mulls, and Kramer, 1983)

recombination: A physical process that can lead to the exchange of segments of polynucleotides between two DNA molecules and which can result in the progeny of a genetic cross possessing combinations of alleles not displayed by either parent. (Bridges and Morgan, 1923)

recombination factor: The gametic output in a hybrid of new combinations of two genes expressed as a fraction or percentage of the total gametic output of new and old combinations combined. The formulas for the gametic output of the double heterozygote AaBb are (a) Coupling: 1/2(1 – p) AB, 1/2p Ab, 1/2p aB, 1/2(1 – p) ab; (b) Repulsion: 1/2p AB, 1/2(1 – p) Ab, 1/2(1 – p) aB, 1/2p ab, where p represents the recombination fraction. Where there is no linkage, $p = 0.5$.

recombination frequency: the proportion of recombinant progeny in the total progeny from a genetic cross.

recombination index: A measure of the average number of chiasmata per nucleus. It is used as an estimate of the meiotic potential for genetic recombination via crossing-over in eukaryotes. Mathematically, recombination index is calculated by adding the gametic chromosome number to the average total number of chiasmata in the mother cell. A high value of recombination index promotes flexibility, whereas low recombination index promotes fitness.

recombination nodules (NODES): Dense objects present on the synaptonemal complex; they could be involved in crossing-over.

recombinational repair: Filling a gap in one strand of duplex DNA by retrieving a homologous single strand from another duplex.

recombinogen: An agent capable of inducing or enhancing recombination between linked genes.

recon: (1) The smallest segment of DNA or subunit of a cistron that is capable of recombination; it may be as small as one deoxyribonucleotide pair; (2) in tissue culture, a viable cell reconstructed by the fusion of a karyoplast with a cytoplast. (Benzer, 1961)

reconstructed cell: *See* RECON (2)

reculture: The process by which a cell monolayer or a plant explant is transferred, without subdivision, into fresh medium. *See* PASSAGE.

recurrent backcrossing: Repetitive sexual crossing of hybrids to one parent in order to eliminate all but the desired characteristic(s) of the donor parent.

recurrent parent: Parent to which hybrid material is crossed in a back cross. *See* NONRECURRENT PARENT.

recurrent selection: A breeding system designed to increase the frequency of favorable genes of quantitatively inherited characteristics by repeated cycles of selection.

redifferentiation: Cell or tissue reversal in differentiation from one type to another type of cell or tissue.

redox gene regulation: A gene regulatory mechanism involving reductional-oxidation of a single conserved cysteine residue in the DNA binding domain of the regulatory proteins.

reduced number of chromosomes: The gametic or haploid number of chromosomes. Some lower organisms have a reduced number in somatic cells, e.g., *Neurospora*. *See* HAPLOID.

reductional division: The first meiotic division. It reduces the number of chromosomes and centromeres to half that of the original cell.

redundant DNA: *See* REPETITIVE DNA.

reduplication hypothesis: The discredited idea that gametes with parental gene combinations undergo a selective mitosis so that their frequencies are greater than the nonparental gene combinations. The reduplication hypothesis was meant to counter the idea of gene combinations generated by crossing-over.

regeneration: In plant cultures, a morphogenetic response to a stimulus that results in the production of organs, embryos, or whole plants.

registered seed: The progeny of breeder or foundation seed and so handled as closely as possible to maintain the genetic identity and purity of a variety. Registered seed is the source of certified seed. Registered seed must be approved and certified by an official seed certification agency.

regulated gene: A gene whose expression is regulated. Its product is synthesized in the cell type, at a time when and in the amount it is required.

regulator element: *See* RECEPTOR ELEMENT.

regulator gene: A gene that codes for a protein, such as a repressor, involved in regulation of the expression of other genes. (Jacob and Monod, 1961a)

regulatory cell: Any cell that controls the fate of other cells.

regulatory RNA: Any of the small RNA molecules that bind to messenger RNA or directly to DNA, or both and regulate gene activity. *See* ANTISENSE RNA.

relative Darwinian fitness: *See* ADAPTIVE VALUE.

relaxed control: A situation in which the numbers of a particular plasmid present in a bacterial cell are not closely regulated by the coupling of plasmid DNA synthesis to host cell DNA synthesis. In relaxed control up to 100 copies of plasmid per cell may be found; this may be further increased using chloramphenicol.

relaxed mutant: A bacterial mutant that does not shut down RNA synthesis abruptly in a depleted growth medium.

relaxed plasmid: A high copy-number plasmid, generally represented by more than 20 copies of a chromosome.

release (termination) factors: Proteins in prokaryotes responsible for termination of translation and release of the newly synthesized polypeptide when a nonsense codon appears in the A-site of the ribosome. Replaced by eRF in eukaryotes. (Ganoza, 1966; Capecchi, 1967)

renaturation: The return of a denatured molecule back to its natural state.

renaturation kinetics: A technique that measures the rate of reassociation or reannealing of complementary single DNA strands derived from a single source that is used to indicate genome size and complexity. *See* COT PLOT.

Renner complexes: In plants like *Oenothera,* this is where all chromosomes are involved in a complex interchange, and genes located on different chromosomes, but close to centromeres, are linked together, forming two gene complexes distributed to different gametes. One gene complex has translocated chromosomes and the other normal chromosomes. These gene complexes are collectively known as Renner complexes. In *O. lamarckiana,* individual complexes are known as *gaudens* and *velans*.

Rep protein: This is an enzyme isolated from certain strains of *E. coli* that opens the strands of the DNA duplex and therefore, acts as a helicase. The enzyme also hydrolyzes ATP.

repair DNA synthesis: Enzymatic excision and replacement of regions of damaged DNA. Repair of thymine dimers by UV irradiation is the best understood example.

repair nuclease: A nuclease that recognizes a DNA strand with a missing base and cleaves out the damaged nucleotide, plus some neighboring nucleotides, allowing later repair by insertion of an undamaged sequence by DNA polymerase and DNA ligase.

repair of DNA: The correction of alterations in DNA structure before they are inherited as mutations.

repeated genes: Some genes may be present in a haploid genome in many copies. These copies may be identical or only similar.

repeating unit: A tandem cluster is the length of the sequence that is repeated; it appears circular on a restriction map.

repetition frequency: The (integral) number of copies of a given sequence present in the haploid genome; it equals 1 for nonrepetitive DNA, greater than 2 for repetitive DNA.

repetitious DNA: DNA in which the base sequences are repeated many times in the genome of multicellular eukaryotes; in highly repetitious DNA, the base sequences may be represented more than 1,000 times. Such DNA is concentrated in heterochromatin, and in chromosomes, it is located in the region of the centromere and telomeres. Moderately repetitious DNA includes the genes for rRNA and tRNA, as well as structural genes coding for the synthesis of histones. Repetitious DNA sequences comprise 20 to 50 percent of most animal genomes and more than 50 percent of many plant genomes.

repetitive and unique sequence hypothesis: Only a small proportion of DNA is unique and is meant for genes carrying the genetic information. The rest of DNA is repetitive and has some other function such as the control of gene activity.

repetitive DNA: DNA comprised of copies of the same nucleotide sequence.

replacement, gene: A method of substituting a cloned gene or part of a gene, which has been mutated *in vitro,* for the wild-type copy of the gene within the host's chromosome.

replacement sites: In a gene, those sites at which mutations alter the amino acid that is coded.

replicase: RNA-dependent RNA polymerase that is able to initiate copying of virus RNA without an added primer. (Spiegelman and Hayashi, 1963)

replicating forms (RF): The structure of a nucleic acid at the time of its replication; the term most frequently used to refer to double-helical intermediates in the replication of single-stranded DNA and RNA viruses.

replication: The process of copying. (1) Replication of DNA; (2) the equal incorporation of all combinations two or more times in an experimental design, which is then said to be replicated.

replication eye: A region in which DNA has been replicated within a longer, unreplicated region.

replication fork: The region of a double-stranded DNA molecule that is being unwound to enable DNA replication to occur.

replication origin: A site on a DNA molecule where unwinding begins in order for replication to occur.

replicative senescence: The permanent termination of DNA replication due to an arrest at G_1/S boundary in interphase.

replicator: A DNA segment that contains an origin of replication and is able to promote the replication of a plasmid DNA molecule in a host cell. *See* AUTONOMOUSLY REPLICATING SEQUENCE.

replicon: A sequentially replicating segment of a nucleic acid controlled by a subsegment known as a replicator. A single replicator is present in the bacterial "chromosome," whereas the chromosomes of eukaryotes bear large numbers of replicons in series. (Jacob and Brenner, 1963)

replisome: The DNA-replicating structure at the Y-junction consisting of two DNA polymerase III enzymes and a primosome (primase and DNA helicase). (Dressler, 1975)

reporter gene: A promoterless gene whose phenotype can be assayed in a transformed organism and that can be used in, for example, deletion analysis of regulatory regions.

repressible control: In this case, the end product (corepressor) acts to repress reproduction of the enzymes. Transcription is initiated in the absence of the end product.

repressible enzyme system: A coordinated group of enzymes, involved in a synthetic pathway (anabolic), is repressible if excess quantities of the end product of the pathway lead to the termination of transcription of the genes for the enzymes. These systems are primarily prokaryotic operons.

repressor: A regulator molecule (protein) produced by a regulator gene that can combine with and repress the action of an associated operator gene. (Pardee, Jacob, and Monod, 1959)

reproductive isolating mechanisms: Environmental, behavioral, mechanical, and physiological barriers that prevent two individuals of different populations from producing viable progeny.

reproductive isolation: The prevention of random mating between one population and the others, or between one species and its relatives.

reproductive success: The relative production of offspring by a particular genotype compared with other genotypes.

repulsion (*trans*-arrangement): An allelic arrangement in which each homologous chromosome has mutant and wild-type alleles. (Bateson, Saunders, and Punnett, 1905)

repulsion linkage: The tendency of two dominant characters, in an F_1 inherited one from one parent and the other from the other parent, to stay apart in the F_2 generation.

res-gene: Resistance gene.

resistance gene interactions: Some pairs of genes for resistance interact to produce a greater level of resistance than either of them individually. Some gene interactions may suppress resistance.

resistance transfer factor (RTF): The segment of a conjugal R plasmid carrying the genetic factors that confer on the plasmid those properties needed for transfer of the plasmid from one cell to the next.

resistant: A characteristic of a host plant such that it is capable of suppressing or retarding the development of a pathogen or other injurious factor.

resolvase: An enzyme actively involved in site-specific recombination between two transposons present as direct repeats in a cointegrate structure.

resolving power: The ability of an experimental technique to distinguish between two genetic conditions (typically discussed when one condition is rare and of particular interest).

restitution: The reunion of the two broken surfaces of a fragmented chromosome, the original structure thereby being restored.

restitution nucleus: A single nucleus found instead of two through the first or second division of meiosis.

restorer gene: A gene, usually dominant, that effectively overcomes the effect of male sterile cytoplasm on male fertility; i.e., it produces functional male gametes even in the presence of the male sterile cytoplasm.

restriction digest: The results of the action of a restriction endonuclease on a DNA sample.

restriction endonuclease: Any type of site-specific endonuclease that recognizes specific DNA sequences and cleaves the DNA duplex at or near these sites to produce DNA fragments (restriction fragments) having discrete molecular weights. These endonucleases are of three types: type I enzymes recognize a specific sequence of DNA but cleave DNA at a site 1000 bp away; type II enzymes cleave DNA within the recognition site; type III are similar to type I, but here recognition and cleavage sites are 5 to 7 bp apart. (Meselson and Yuan, 1968)

restriction enzyme: One of a number of enzymes that break DNA molecules down by causing cleavage at specific points in the molecule; the points are determined by the base sequence. (Linn and Arber, 1968)

restriction fragment length polymorphism (RFLP): Natural variations in the banding patterns of electrophoresed restriction digests, due to variation in the length of restriction fragments. RFLPs have certain traits—lack of dominance, multiple allelic forms, lack of pleiotropic effects on agronomic traits, codominance, no measurable effect on phenotype, and no effect of environment on its occurrence (Grodzickev et al., 1974). Botstein and coworkers (1980) were the first to propose its use in construction of linkage maps.

restriction fragments: Products of the complete digestion of a chosen target DNA with a restriction endonuclease. Fragments are flanked by the specific recognition sites for that restriction endonuclease.

restriction map: A physical map of a piece of DNA showing recognition sites of specific restriction endonucleases separated by lengths marked in number of bases. Also known as *cleavage map*. (Danna and Nathans, 1971)

restriction site: The base sequence at which a restriction endonuclease cuts the DNA molecule, usually a point of symmetry within a palindromic sequence.

restriction site mapping: A technique in which genomic DNA is cleaved with restriction endonuclease into hundreds of fragments of varying size. The fragments are separated by gel electrophoresis and blot transferred to a cellulose nitrite film.

retroelement: Any of the integrated retroviruses or the transposable elements that resemble them.

retroposon: *See* RETROTRANSPOSON. (Rogers 1983)

retroregulation: The ability of a sequence downstream to regulate translation of an mRNA. (Schindler and Echols, 1981)

retrotransposon: A transposon that mobilizes via an RNA form; the DNA element is transcribed into RNA, and then reverse-transcribed into DNA, which is inserted at a new site in the genome. (Bocke et al., 1985)

retrovirus: Any of a family of eukaryotic, single-stranded RNA viruses that replicate by way of a duplex DNA provirus integrated into cellular DNA and synthesized by reverse transcription. (Fenner, 1976)

reversal of central dogma: A phenomenon mostly observed in RNA viruses where reverse transcriptase uses RNA as a template to produce complementary DNA.

reverse genetics: A type of genetic analysis in which a cloned gene is altered at a predetermined site by site-specific mutagenesis and then assayed for changes in gene function. It is the opposite of conventional genetics where a particular mutant is selected first, and then its genotype is explored. (Weisman et al., 1979; Mantei, Boll, and Weissman, 1979)

reverse mutation: Any change toward the standard (wild-type) by way of a second mutational event. Also known as *reversion* or *back mutation*.

reverse self-splicing: In group I introns, the full reversibility of the self-splicing reaction *in vitro*. It has implications for transposition of group I introns. (Woodson and Cech, 1989)

reverse tandom duplication: This occurs when the duplicate segment of a chromosome is present adjacent to original sequences but in reverse order.

reverse transcriptase: An enzyme that synthesizes a DNA copy on an RNA template. (Baltimore, 1970; Temin and Mizutani, 1970)

reverse transcription: The synthesis of a DNA copy on an RNA template.

reverse translation: A technique for isolating genes (or mRNAs) by their ability to hybridize with a short oligonucleotide sequence prepared by predicting the nucleic acid sequence from the known protein sequence.

reversion: Reversion of mutation is a change in DNA that either reverses the original alteration (true reversion) or compensates for it (second site reversion in the same gene).

revertants: Revertants are derived by reversion of a mutant cell or organism.

RFLP fingerprinting: This technique of obtaining fingerprints is based on restriction fragment length polymorphism by introducing oligonucleotides complementary to microsatellites (e.g., [GATA]$_4$) as end-labeling probes.

RFLP markers: Any marker resulting from changes in the length of genomic DNA with specific restriction endonucleases.

rho factor: A protein required to halt transcription of certain genes. (Roberts, 1969)

rho-dependent terminator: A DNA sequence signaling the termination of transcription; termination requires the presence of rho protein.

rho-independent terminator: A DNA sequence signaling the termination of transcription; termination does not require the presence of rho protein.

ribonuclease: An enzyme that degrades RNA.

ribonucleic acid (RNA): A single-stranded nucleic acid molecule, synthesized principally in the nucleus from deoxyribonucleic acid, composed of a ribose-phosphate backbone with purines (adenine and guanine) and pyrimidines (uracil and cytosine) attached to the sugar ribose. RNA is one of several kinds of functions used to carry the "genetic message" from nuclear DNA to the ribosomes.

ribonucleoprotein (RNP): A complex of RNA and protein. Particles of RNA occur in both nuclei and cytoplasm of eukaryotic cells and are formed of eukaryotic cells. They are formed in order to protect RNA from degradation and probably also to assist in transport and localization within the cell.

ribonucleoside: The portion of an RNA molecule composed of one ribose molecule plus either a purine or a pyrimidine.

ribonucleotide: The portion of an RNA molecule composed of one ribosephosphate unit plus a purine or a pyrimidine.

ribosomal cistron: The gene coding for ribosomal RNA.

ribosomal DNA (rDNA): DNA whose transcripts are processed into the RNA components of the ribosomes.

ribosomal RNA (rRNA): The RNA molecules that act as structural components of the ribosomes. (Kurland, 1960)

ribosome: One of the protein-RNA structures on which translation occurs. Ribosomes are composed of a small and a large subunit. The intact prokaryotic ribosome particle is 70S, which contains the 30S and 50S subunits. The 30S subunit consists of one 16S rRNA molecule plus 21 different proteins. The 50S subunit is constituted by one 5S rRNA molecule plus one 23S rRNA molecule plus 32 different proteins. 5S rRNA contains 120 nucleotides; 16S rRNA has 1,541 nucleotides; 23S rRNA has 2,904 nucelotides. Eukaryotic ribosome is larger (80S) than prokaryotic ribosome. The two eukaryotic subunits are of size 40S and 60S. The 40S subunit has one 18S rRNA molecule plus about 30 proteins. The 60S subunit contains one 5S rRNA plus one 5.8S rRNA plus one 28S rRNA molecule plus about 50 proteins. (Roberts, 1958; Dintzis, Borsook and Vinograd, 1958)

ribosome binding site: The nucleotide sequence that acts as the site for attachment of a ribosome to an mRNA molecule. Also known as *Shine-Dalgarno* or *S-D sequence. See* SHINE-DALGARNO SEQUENCE. (Shine and Dalgarno, 1974)

ribosome biosynthesis: The assembly of ribosomal particles from RNA and protein components. This is coordinated in such a way in eukaryotes and prokaryotes that neither excess protein nor nucleic acids accumulate.

ribozyme: An RNA molecule that possesses catalytic activity.

rifampicin (rifampin): An antibiotic that inhibits RNA synthesis. Rifampicin, a derivative of rifamycin SV, specifically inhibits DNA-dependent bacteria and chloroplast, but not mammalian RNA polymerase. It is also reported to inhibit the assembly of DNA and protein into mature virus particles. Also known as *rifamycin*.

right splicing junction: The boundary between the right end of an intron and the left end of the adjacent exon.

ring chromosomes: A physically circular chromosome usually found in bacteria.

R-line: The pollen parent line containing fertility-restoring genes, crossed with the A-line in the production of hybrid seed.

R-line (restorer): The pollen parents used in hybrid seed production. They restore fertility of the A-line, i.e., the male sterile female line; the

B-line or maintainer, on the other hand, maintains the male sterility of A-line on crossing, i.e., A × B→A, since B is isogenic to A.

R-looping: The technique in which an RNA molecule is annealed to the complementary strand of a partially denatured DNA molecule—the formation of the RNA. The DNA hybrid displaces the opposite DNA strand as a single-stranded bubble. These R-loops can be visualized under the electron microscope using the Kleinschmidt technique. It was the R-looping technique that first revealed the presence of introns in eukaryotic genes. *See* D-LOOP.

RNA: Ribonucleic acid—one of the two forms of nucleic acid in living cells.

RNA editing: RNA editing results in changes in the nucleotide sequence of mitochondrial transcripts such that the RNA sequence differs from the DNA template from which it is transcribed. In this process uridine (U) residues are added or deleted at multiple precise sites. RNA editing also involves C-to-U change, multiple C-to-U or U-to-C changes, addition of G or C residues, A-to-G modification, and conversion of U-to-A, U-to-G, and A-to-G. RNA editing takes place with the help of guide RNAs (gRNAs). mRNAs are edited most frequently, whereas edits of tRNAs and rRNAs are extremely rare.

RNA folding: In the folding of highly structured RNA molecules, a model proposes that formation of one helical domain precedes that of a second helical domain. (Zarrinker and Williamson, 1994)

RNA ligase: An enzyme that can join RNA molecules together. (Silber, Malathi, and Hurwitz, 1972)

RNA plasmid: Unusual genetic elements comprised of single or double-stranded RNAs found in plant mitochondria of some male sterile lines. (Finnegan and Brown, 1986)

RNA polymerase: An enzyme capable of synthesizing an RNA copy of a DNA template.

RNA polymerase core enzyme: In bacteria, it consists of 5 subunits ($\alpha_2\beta\beta'\omega$) and is capable of elongating already initiated transcription by holoenzyme.

RNA polymerase holoenzyme: The version of the *E. coli* RNA polymerase that has the subunit composition $\alpha_2\beta\beta'\omega\sigma$. It is involved in efficient recognition of promoter sequences and is capable of initiation of transcription.

RNA primer (pRNA): RNA that mediates DNA replication. It defines a starting point and an RNA-DNA transition point. (Zechel, Bouché, and Koranberg, 1975)

RNA processing: This involves the steps that heterogenous nuclear RNA (pre-mRNA) transcribed by RNA polymerase II undergoes to produce a finished mRNA molecule. These steps include: (a) addition of a "cap" (5-m^7Gppp) at the 5′ end, (b) addition of a poly(A) tail at the 3′ end, and (c) splicing of noncoding sequences (introns) and methylation of one out of every 400 adenines present.

RNA replicase: A polymerase enzyme that catalyzes the self-replication of single-stranded RNA.

RNA self-splicing: A set of linked cleavage-ligation reactions mediated by RNA itself and conserving the number of phosphodiester bonds. This process is observed in some nuclear preribosomal RNAs and various mitochondrial chloroplast pre-mRNAs. (Cech, Zaug, and Grabowski, 1981)

RNA sequence analysis: The determination of the base sequence in RNA using enzymes in a method directly analogous to the Maxam-Gilbert DNA technique, except that base-specific ribonucleases are employed which cleave specifically next to G, A, or A and U and are specific for pyrimidines.

RNA splicing: The removal of large noncoding sequences (introns) from the primary RNA transcript, followed by rejoining of the nonadjacent coding sequences (exons) to produce the functional mRNA. The basic mechanism of RNA splicing seems to involve a series of transesterifications. (Berget, Moore, and Sharp, 1977)

RNA transcript: An RNA copy of a gene.

RNA transcriptase: The enzyme responsible for transcribing the information encoded in DNA into RNA; also called transcriptase or RNA polymerase.

RNA trimming: The removal of stretches of RNA at one or both ends of primary RNA transcript.

RNA-binding proteins: There are a multitude of RNA-binding proteins that play a role in posttranscriptional regulation of gene expression. These proteins contain RNA-binding motifs. (Dreyfuss, Phillipson, and Mattaj, 1988; Dreyfuss et al., 1993; Burd and Dreyfuss, 1994)

RNA-coding triplets: Triplets (codons) found on mRNA that code for the amino acids—the code being recognized by the anticodon present on the molecule of tRNA charged with the specific amino acid. *See* GENETIC CODE.

RNA-dependent DNA polymerase: A group of enzymes that catalyze the formation of DNA molecules from RNA templates. These occur in some viruses (e.g., those that produce tumors). *See* REVERSE TRANSCRIPTASE.

RNA-driven hybridization reactions: The reaction that uses an excess of RNA to react with all complementary sequences in a single-stranded preparation of DNA.

RNase: An enzyme that hydrolyzes RNA.

RNase D: The enzyme responsible for processing pre-tRNA by cleaving at the 3′ termini of the mature tRNA sequences.

RNase H: An enzyme that degrades specifically the RNA strand of RNA-DNA molecules.

RNase mapping: A method used to detect and quantify specific RNAs and to determine structural features of RNA molecules by hybridization with SP6 single-stranded ^{32}P-RNA probes. An excess of ^{32}P-RNA complementary to the test RNA is synthesized with SP6 RNA polymerase and hybridized to the RNA sample. Ribonuclease treatment digests the unhybridized single-stranded ^{32}p-probe. ^{32}P-RNA-RNA hybrids are ribonuclease resistant and can be detected and quantified by gel electrophoresis.

RNase P: An enzyme that removes precursor sequences from the 5′ end of the pre-tRNA in eubacteria.

RNA-transcript mapping: A procedure involving hybridization of the RNA to be analyzed with a radioactivity labeled and purified DNA probe under the conditions that favor RNA-DNA or DNA-DNA duplexes. After digestion with S1 nuclease, the nuclease resistant fragments are resolved through polyacrylamide gel electrophoresis.

Robertsonian rearrangements: Fusions and fissions of chromosomes.

Robertsonian translocations: The joining together of two acrocentric chromosomes at the centromeres.

rolling circle replication: A type of DNA replication that accounts for a circular DNA molecule producing linear daughter double helixes. (Gilbert and Dressler, 1968)

rot value: The product of RNA concentration and time of incubation in an RNA-driven hybridization reaction.

rough endoplasmic reticulum (RER): Consists of endoplasmic reticulum associated with ribosomes.

R′ plasmid: A plasmid carrying selectable bacterial genes.

rubisco: A bifunctional 1,5-ribulose bisphosphate carboxylase/oxygenase enzyme that catalyzes both photosynthesis (carboxylation) and photorespiration (oxygenation) processes simultaneously in chlorophyll-bearing plants.

runaway plasmid vector: A plasmid vector, which at a low temperature (30°C), is present in a moderate number of copies per cell because of some control on plasmid replication, but at a higher temperature (35°C), all control on plasmid replication is lost and the number of copies per cell increases continuously. (Uhlin and Clark, 1981)

runaway replication: Uncontrolled replication of plasmid molecules that multiply within the host producing up to several thousand copies. This usually halts cell division but is a very powerful way of amplifying a gene product. Plasmid vectors that are temperature sensitive for a copy number control element have been constructed as runway replication vectors.

S phase: *See* SYNTHETIC PHASE. (Howard and Pelc, 1953)

S value: The unit of measurement of a sedimentation coefficient. *See* SEDIMENTATION COEFFICIENT.

S_0: Symbol used to designate the original selfed (self-pollinated) plant.

S_1, S_2, etc: Symbols for designating first selfed generation (progeny of S_0 plant), second selfed generation (progeny of S_1 plant), etc., of the original self-pollinated plant.

S_1 endonuclease: An enzyme that specifically degrades unpaired, single-stranded DNA sequences.

S_1 nuclease: An enzyme that degrades specifically single-stranded molecules or single-stranded regions in predominantly double-stranded nucleic acid molecules.

Sal I box: A sequence (AGGTCGACCAGATNTCCG) that directs transcription termination of ribosomal DNA (mouse) transcription by RNA polymerase I. This sequence is present eight times in the spacer region downstream of the $3'$ end.

S-allele: Any allele of those genes that control incompatibility in plants. Alleles present in both style and pollen are referred to as "matched S-alleles." S_F-alleles are alleles that belong to same series of multiple alleles as S-alleles but which control self-fertility.

saltation hypothesis: A hypothesis that explains evolution by step-wise changes in a population.

saltatory DNA replication: A sudden lateral amplification to produce a large number of copies of some DNA sequence. (Britten and Davidson, 1971)

satellite: A portion of the chromosome separated from the main body of the chromosome by secondary constrictions. (Navashin, 1912)

satellite association: All the chromosomes that have satellites tend to arrange themselves in a group during metaphase. This poses difficulty in assigning a particular number to a particular chromosome.

satellite (SAT) chromosome: A chromosome with a secondary constriction that separates a satellite from the rest of the chromosome. (Heitz, 1931)

satellite DNA: DNA comprising clustered repetitive sequences, so-called because it forms a satellite band in a density gradient. (Sueoka, 1961; Kit, 1961; Britten and Davidson, 1971)

satellite RNA: An RNA that is transmitted along with the virus but does not infect or damage the host.

saturation density: The density to which cultured eukaryotic cells grow *in vitro* before division is inhibited by cell-cell contacts. Thus, this is the maximum cell number attainable under specified culture conditions and is expressed as the number of cells per square centimeter in a monolayer culture or the number of cells per cubic centimeter in a suspension culture.

saturation hybridization: The experiment that has a large excess of one component, causing all complementary sequences in the other component to enter a duplex form.

scaffold: The eukaryotic chromosome structure remaining when DNA and histones have been removed; it is made from nonhistone proteins.

scanning: A proposed system for the initiation of eukaryotic translation in which the ribosome attaches to the 5′ terminal-cap structure of the mRNA and then scans along the molecule until it reaches an initiation codon.

scanning hypothesis: Proposed mechanism by which the eukaryotic ribosome recognizes the initiation region of an mRNA after binding the 5′ capped end. The ribosome scans the mRNA for the initiation codon.

scarce (complex) mRNA: mRNA that consists of a large number of individual mRNA species, each present in very few copies per cell.

scn DNA: Single copy nuclear DNA.

screening: A process whereby every colony is tested individually for a particular property.

screening technique: A technique used to determine the genotype or phenotype of some property of an organism.

scRNA: Any one of several small cytoplasmic RNAs; molecules present in the cytoplasm and (sometimes) nucleus.

scRNPs: Small cytoplasmic ribonucleoproteins that are formed when small cytoplasmic RNAs (scRNAs) are associated with proteins. scRNPs are part of signal recognition particles (SRPs) bound to endoplasmic reticulum. *See* SIGNAL RECOGNITION PARTICLE.

second messenger: Any agent responsible for the transmission of hormonal stimuli within the cell. cAMP is a secondary messenger with tissue-specific action and a variety of different effects in different organs and organisms. It activates various protein kinases which in turn are able to regulate the activities of numerous other enzymes. cAMP also influences the negatively controlled activities of genes (catabolic repression). (Suderland, 1972)

second site reversion: A second mutation that reverses the effect of a previous mutation in the same gene, although without restoring the original nucleotide sequence. (Yanofsky, Helinski, and Maling, 1961)

secondary constriction: A noncentric constriction, containing a nucleolar organizer region (NOR).

secondary protein structure: The flat or helical configuration of the polypeptide backbone of a protein.

secondary trisomic: A trisomic having an additional chromosome as an isochromosome for one of the chromosome arms of the complement (Blakeslee, 1921). Also called *monoisotrisomic* by Kimber and Sears (1968).

second-division segregation (SDS): The segregation of a heterozygous pair of alleles at a locus at meiosis II. SDS produces a 1:1:1:1 ratio during tetrad analysis.

sector: An area of tissue whose phenotype is detectably different from the surrounding tissue phenotype.

sedimentation coefficient: A value used to express the velocity with which a molecule or structure sediments when centrifuged in a dense solution.

seed: A mature ovule with its normal covering. A seed consists of the seed coat, embryo, and in certain plants, an endosperm.

seeding efficiency: *See* ATTACHMENT EFFICIENCY.

seedling resistance genes: The genes that confer resistance against a pathogen to a host at seedling stage.

segmental allopolyploid: An allopolyploid in which the combined genomes are homologous in many small segments throughout the complement; crossing-over may recombine material from different genomes. (Stebbins, 1947)

segmentation gene: In eukaryotes, any of the genes that determine the number and polarity of body segments. *See* HOMEOBOX.

segregating population: The appearance of more than one genotype in the progeny of a heterozygote.

segregation: The separation of paternal and maternal chromosomes from each other at meiosis; the separation of alleles from each other in heterozygotes; and the occurrence of different phenotypes among offspring, resulting from chromosome or allele separation in their heterozygous parents. Mendel's first principle of inheritance.

segregation distortion: Any systematic deviation from equal representation of alleles or homologous chromosomes among the functional gametes. It leads to meiotic drive. (Sandler and Hiraizumi, 1961)

segregation, principle of: The separation of homologous chromosomes, or members of allele pairs, into different gametes during meiosis.

segregation ratio: The expected (Mendelian) or observed ratio between genotypes or phenotypes in the progeny of a cross.

segregational load: Genetic load caused when population is segregating less-fit homozygotes because of heterozygote advantage.

selection: Differential (nonrandom) reproduction of different genotypes. (Darwin, 1859)

selection intensity (i): The amount of selection pressure applied, expressed as the proportion of the population selected. The larger the size of i, the more stringent is the selection pressure (i.e., a small fraction is selected). i = 0 to 2.64, correspondingly, means 100 to 1 percent proportion selected. The measure of selection intending to alter the frequency of a gene in a given population is known as *selection pressure.* (Wright, 1921)

selection limit: Fixation of all favorable alleles in the population, resulting in the exhaustion of genetic variance hence, no further response to selection.

selection regime: The particular set of relative fitness for alternative genotype within a population living in a particular environment.

selection response (R): In plant and animal breeding, the difference between the mean of the individuals selected to be parents and the mean of their offspring.

selection-mutation equilibrium: An equilibrium allelic frequency resulting from the balance between selection against an allele and mutation recreating this allele.

selective advantage: The genotypic condition of a cell or an individual, or genetic class of individuals that increases its chances, relative to others, of representation in later generations of cells or individuals.

selective mating: Nonrandom mating. Phenotypic differences influence the separation of the F_2 progeny into distinct types.

selective neutrality: A situation in which different alleles of a certain gene confer equal fitness.

selective system: An experimental technique that enhances the recovery of specific (usually rare) genotypes.

self-assembly: The ability of certain multimeric biological structures to assemble from their component parts through random movements of the molecules and formation of weak chemical bonds between surfaces with complementary shapes.

self-incompatibility: Failure of pollination and/or fertilization even though both male and female organs in a hermaphrodite flower are fully functional. (Stout, 1917)

self-incompatibility alleles: These are alleles of a multiple allelic series that determine compatibility of perfectly fertile male and female flowers. If a pollen grain bears a self-incompatibility (S) allele that is also present in the maternal parent, then it will not germinate. But if that allele is not present in the maternal tissue, then the pollen grain produces a pollen tube containing the male nucleus, and this tube affects fertilization.

selfish DNA: In eukaryotes, any tandemly repeated for, dispersed repetitive DNA sequences that appear to have no function and apparently contribute nothing to the cell in which they are found. Also known as *ignorant DNA* or *junk DNA*. (Dover and Ford-Doolite, 1980)

self-replicating peptide: A peptide capable of templating its own synthesis autocatalytically. For example, a 32-residue 2-helical peptide based on the leucine-zipper domain of yeast transcriptional factor GCN4 can act as a self-replicating peptide by amide-bond condensation. (Lee et al., 1996)

self-sterile: Failure to complete fertilization and seed set after self-pollination.

self-sterility: The incapability of producing seed when self-pollinated. Several alleles (S_1, S_2, S_3, etc.) are responsible for this phenomenon. (Darwin, 1877)

semiconservative DNA replication: The mode of DNA replication in which each daughter double helix comprises one polynucleotide from the parent and one newly synthesized polynucleotide. (Delbrük and Stent, 1957)

semidiscontinuous replication: The mode of DNA replication in which one new strand is synthesized continuously while the other is synthesized discontinuously.

semidominance: *See* DOMINANCE.

semigamy: The male gamete penetrates into the egg cell and segregation of maternal and paternal nuclei takes place. There is no fertilization. (Battaglia, 1963)

semilethal: Partially, but not completely, lethal. Semilethality can be either gametic or zygotic. (Muller and Altenburg, 1919)

semilethal mutation: *See* VITALITY MUTATION.

semispecies: The populations that have acquired some but not yet all the attributes of a species rank. These are border cases between species and subspecies. (Mayr, 1940)

semisterility: The nonviability of a proportion of gametes or zygotes.

sense codon: A codon specifying a particular amino acid in protein synthesis. Also known as *sense word*.

sense DNA: The strand of DNA that is used as a template during transcription.

sense RNA: The RNA formed on the template or antisense strand of the gene. Also known as *mRNA*. *See* MESSENGER RNA.

sense strand: The one strand of the two DNA strands composing a gene that does not act as a template for the formation of a transcript. This strand has the same sense as the mRNA with Us in the RNA in place of Ts.

sensitive development period: The period of development during which the action of a gene is sensitive to the influence of external conditions.

sensor gene: A hypothetical eukaryotic regulatory gene that is sensitive to external, developmental signals. (Britten and Davidson, 1971)

sensor region: In the Britten-Davidson (1969) model of eukaryote gene regulation, the sensor region responds to a signal in the environment (perhaps a hormone) and activates integrator genes, which activate receptor genes, which activate structural genes. (Britten and Davidson, 1969)

sequence analysis: The determination and study of the sequence of bases in DNA or of amino acids in proteins.

sequence map: A pictorial representation of the sequence of amino acids in a peptide or of bases in a oligonucleotide.

sequence-tagged microsatellite sites (STMS) analysis: A strategy used for introducing microsatellite amplification by locus specific polymerase chain reaction (PCR). This strategy involves screening a genomic library for microsatellite repeats, the sequencing of positive clones, designing primers that flank the repeats, and amplifying genomic DNA with these locus-specific primers in a PCR with genomic DNA. This strategy is known as *microsatellite-primed PCR (MP-PCR)*, which yields multilocus banding patterns resembling those obtained by *random amplified polymorphic DNA (RAPD)*. This is the most popular way to exploit microsatellite polymorphism for genetic mapping and genotype identification.

sequence-tagged sites (STS): Polymerase chain reaction (PCR)–based, restriction fragment length polymorphism (RFLP) that uses the end sequence of a genomic clone as a primer.

sequencing gel: A long, polyacrylamide salt gel that has sufficient resolving power to separate single-stranded fragments of DNA or RNA which differ in length by only a single nucleotide. Electrophoresis is carried out at high voltage and with the gel in a vertical position.

serial analysis of gene expression (SAGE): A technique that allows quantitative and simultaneous analysis of a large number of transcripts. It is based on the principles that a short nucleotide tag (9-10 bp) contains sufficient information to uniquely identify a transcript, and that concentration of short sequence tags allows efficient analysis of transcripts in a serial manner. (Velculescu et al., 1995)

sex chromosome: A chromosome that is involved in sex determination. These chromosomes are not similarly distributed in both sexes. Distribution of a sex chromosome to one but not the other of the products of meiosis determines differences in sex of the offspring.

sex determination: Mechanisms responsible for determining the sex of a zygote during development.

sex factor: A bacterial episome (e.g., the F plasmid in *E. coli*) that enables the cell to be a donor of genetic material. The sex factor may be propagated in the cytoplasm, or it may be integrated into the bacterial chromosome.

sexduction: The incorporation of bacterial genes into F factors and their subsequent transfer by conjugation to a recipient cell. Also known as *F-duction* or *F-mediated transduction.* (Jacob et al., 1960)

shear: (1) It is the movement of one layer relative to a parallel adjacent layer; (2) to fragment DNA molecules into smaller pieces. DNA is a very susceptible to hydrodynamic shear forces. Forcing a DNA solution through a hypodermic needle will fragment it into smaller pieces. The size of all the fragments obtained is inversely proportional to the diameter of the needle's bore. The actual sites at which the shear force breaks a DNA molecule are approximated by random shear and then cloned (by either tailing their ends or using linkers), so as to create a complete gene library of an organism.

shift: A chromosome aberration in which a chromosome segment is transposed to a new location in the same or a different chromosome. The shift is a sample type of translocation.

shifty transfer RNA: A normal tRNA that promotes frameshifting at a site of nontriplet movement called a shifty codon. Such tRNAs can cause two-base and four-base translocation errors by mistranslating certain noncognate codons. (Weiss, 1984)

Shine-Dalgarno sequence: The prokaryotic ribosome-binding site (AGGAGG) located on mRNA just prior to the AUG initiation codon that has complementarity with the 3′ end of 16S rRNA. Also known as *S-D sequence* or *ribosome binding site.* (Shine and Dalgarno, 1974)

shoot apex (TIP) culture: A structure consisting of the shoot apical meristem plus one to several primordial leaves, usually measuring from 0.1 to 1.0 mm in length; in instances where more mature leaves are included, the structure can measure up to several centimeters in length.

short interspersed nucleotide element (SINE): Any of the short (70 to 300 bp) repetitive sequence elements interspersed with longer single-copy regions in eukaryotic genomes. These have well-defined ends and most have a functional internal promoter. Most of these are retroposons. The best characterized family of SINEs is the mammalian Alu-like sequence family. (Davidson et al., 1973)

short-period interspersion: A pattern in a genome in which moderately repetitive DNA sequences of ~300 bp alternate with nonrepetitive sequences of ~1000 bp.

short-term gene regulation: Gene regulation recognized in eukaryotes that operates in response to fluctuations in environment, changes in activities or concentrations of substrates, and end products or hormone levels. It is a feature of both developing and fully differentiated cells, or operates even when the cell is undergoing differentiation.

shotgun experiment: The cloning of an entire genome in the form of randomly generated fragments to create a gene library for the species that is available for later studies.

shuttle mutagenesis: In this process of mutagenesis, first a gene is cloned and then mutated through the use of transposons, and the mutated gene is then transferred back into the original organism. (Seifert et al., 1986)

shuttle vector: A plasmid constructed to have origins for replication for two hosts (for example, *E. coli* and *S. cerevisiae*) so that it can be used to carry a foreign sequence in either prokaryotes or eukaryotes.

SI system: System International d′ Unites. An international system of units for the measurements of length, volume, and mass. It is based on the metric system.

SI mapping: A method for determining the transcriptional organization of a gene.

sibling species: Morphologically similar or identical populations that are reproductively isolated. These species may be sympatric or allopatric. These species may exit at the same time or at different times. (Mayr, 1942)

sigma chain: One of the polypeptide chains composing RNA polymerase, which is essential for the recognition of signals in the DNA for the start of transcription.

sigma factor: A proteinous component of RNA polymerase that is necessary for initiation of transcription. It is not required for elongation of the transcript. (Burgess et al., 1969; Bautz, Bautz, and Dunn, 1969)

signal codons: The codons that code for either start or stop signals in protein synthesis.

signal hypothesis: The major mechanism whereby proteins that must insert into or across a membrane are synthesized by a membrane-bound ribosome. The first thirteen to thirty-six amino acids synthesized, termed a signal peptide, are recognized by a signal recognition particle that draws the ribosome to the membrane surface. The signal peptide may be removed later from the protein.

signal molecule: Any molecule capable of indicating to living cells to perform a specific biological function. Such a signaling phenomenon has been observed in *Agrobacterium*.

signal peptide: A stretch of predominately hydrophobic amino acids that directs the polysomes to the endoplasmic reticulum during translation. *See* SIGNAL HYPOTHESIS. (Blobel and Dobberstein, 1975)

signal recognition particle (SRP): A ribonucleoprotein particle consisting of seven RNAs and six different proteins. It helps in the targeting of proteins into the endoplasmic reticulum. *See* SIGNAL HYPOTHESIS. (Walter and Blobel, 1980; Singer, 1982)

signal sequence: The region of a protein (usually N-terminal) responsible for cotranslational insertion into membranes of the endoplasmic reticulum.

signal transducers and activators of transcription (STATs): These are the proteinous factors that after phosphorylation by kinases, move to the nucleus, bind to specific DNA elements, and direct transcription of interferon genes whose protein products interfere with viral replication.

signal transduction: The process by which a receptor interacts with a ligand at the surface of the cell and then transmits a signal to trigger a pathway within the cell. Also known as *cell signaling.*

silencer sequence: DNA sequence that helps to inhibit transcription, located at a long distance from the core promoter.

silent alleles: *See* AMORPHIC ALLELES.

silent gene: A gene without a promotor that can be activated by an insertion sequence placed near the unexpressed coding sequence. The insertion sequence provides a promotor that directs transcription into sequences adjacent to its insertion site.

silent mutation: An alteration in a DNA sequence that does not affect the expression or functioning of any gene or gene product. (Sonneborn, 1965)

silent sites: The sites in a gene described as those positions at which mutations do not alter the product.

simple gene: A continuous sequence in a nucleic acid that specifies a particular polypeptide or functional RNA.

simple sequence DNA: *See* SATELLITE DNA.

simple sequence repeats (SSRs): *See* MICROSATELLITES.

simple sequences (SSs): *See* MICROSATELLITES.

simple tandem repeats (STRs): *See* MICROSATELLITES.

simple transcription unit: When a primary transcript produces only one type of mRNA.

single cross: (1) A cross between two inbred lines, or pure-line varieties (in self-pollinated species); (2) the progeny of a cross between two inbred lines or pure-line varieties.

single primer amplification reaction (SPAR): A variant of MP-PCR technique that involves single di-, tri-, tetra-, and pentanucleotide repeats to amplify genomic DNA across a panel of eukaryotes.

single-seed-descent: Selection procedure in which F_2 plants and their progenies are advanced by single seeds until genetic purity is virtually attained.

single-strand assimilation: The ability of RecA protein to cause a single strand of DNA to displace its homologous strand in a duplex; that is, the single strand is assimilated into the duplex.

single-strand binding (ssb) protein: One of the proteins that attaches to single-stranded DNA in the region of the replication fork, preventing reannealing of unreplicated DNA.

single-strand exchange: A reaction in which one of the strands of a duplex of DNA leaves its former partner and instead pairs with the complementary strand in another molecule, displacing its homolog in the second duplex.

single-stranded DNA: A form of DNA that is not base paired with a second strand of DNA or RNA and thus, can hybridize with other suitable polynucleotides.

sister chromatid: Copies of a chromosome produced by its replication and joined together by a centromere.

sister chromatid differential (SCD) staining: Differential staining with fluorochromes and conventional dyes of sister chromatids, when the DNA has been substituted for two rounds of replication with 5′ bromodeoxyuridine or for one round followed by another in thymidine. (Latt, 1973)

sister chromatid exchange (SCE): The exchange, by breakage and reunion, of DNA sequences between sister chromatid and apparently homologous sites. SCE occurs during DNA synthesis between DNA double strands having the same polarity. Analysis of SCEs is a sensitive and reproducible indicator for DNA damage induced by mutagens and carcinogens. (Taylor, Woods, and Hughes, 1957)

site: The position occupied by a mutation within the gene. (Demerc, 1956)

site-directed mutagenesis: The construction of mutations at the predetermined site of a cloned DNA, precisely defining the nature of mutational change and then testing the functional effect of that mutation *in vivo* or *in vitro*. Also known as *site-specific mutagenesis* or *directed mutagenesis*. (Hutchinson, Phillips, and Edgell, 1978)

site-specific mutation: A technique in which a cloned gene is specifically mutated *in vitro* and then used to replace the wild-type copy of the gene in the donor organism. The technique has two advantages over conventional mutagenesis procedures: (a) the proportion of desired mutants obtained is high—up to 50 percent of the viable cells are mutant in some methods; and (b) no other genes in the organism are mutated and, therefore, all the desirable characteristics of the strain are retained.

site-specific recombination: Any system of genetic recombination that promotes genetic exchange between specific DNA sequences. It plays a crucial role in DNA transposition, gene regulation, and generation of genetic diversity.

slow component: A component of a reassociation reaction that is the last to reassociate; it usually consists of nonrepetitive DNA.

slow rusting: Involves host-pathogen combinations in which rust develops slowly but never reaches a high degree of severity. Slow rusting is a complex trait and can be resolved into several components—reduced frequency of penetration, increased latent period, reduced number of uredia per unit area of host surface, reduced pustule size, reduced pustule expansion, reduced sporulation, reduced spore deposition, and high infection threshold—that may act independently or complementarily to arrest fungal infection. *Incomplete resistance* or *partial resistance* are considered to be the cause of slow rusting and are used synonymously with slow rusting. (Caldwell, Roberts, and Eyal, 1970)

Sma I: A type II restriction enzyme from the bacterium *Serratia macesccens* that recognizes the DNA sequences shown below and cuts at the sites indicated by the arrows.

```
               ↓
5′  C  C  C  G  G  G  3′
3′  G  G  G  C  C  C  5′
               ↑
```

small nuclear ribonucleoproteins (snRNPs): Small nuclear ribonucleoproteins are components of the spliceosome—the intron-removing apparatus in eukaryotic nuclei.

small nuclear RNA (snRNA): The RNA component of the nucleus that comprises relatively small molecules thought to be involved in splicing and other transcript-processing events. *See* U1 TO U11 snRNAS. (Weinberg and Penman, 1968)

small RNA chaperons: Small U-RNA molecules involved in ribosome biosynthesis. (Steitz and Tycowski, 1995)

smallest known gene: Microcin C7 (Mcc C7), a modified linear heptapeptide that inhibits protein synthesis in Enterobacteriacae, consists of Acetyl Met-Arg-Thr-Gly-Asn-Ala-Asp. It is synthesized from a 21 bp open-reading-frame gene called mcc A. This is one of the smallest genes so far identified. (Jose, Pastor, and SanMillan, 1994)

smart transcription factors: These are the proteinous transcription factors that are smart enough to make their contact with proteins through their phosphorylation to initiate transcription.

S-9 mix: A liver-derived supernatant used in the Ames test to activate or inactivate mutagens.

solenoid: A higher-order structure in a eukaryotic chromosome in which 100 Å fiber of nucleosomes are stacked with 5 or 6 nucleosomes per helix. H1 histone helps in solenoid formation that produces a package ratio of 1:42. (Finch and Klug, 1976)

soluble RNA (sRNA): *See* TRANSFER RNA (tRNA).

soma: The body cells of mammals and flowering plants that normally have two sites of chromosomes, one derived from each parent. (Weismann, 1885)

somaclonal variation: Phenotypic variation, either genetic or epigenetic, displayed among somaclones, regenerated from tissue culture involving callus formation. (Larkin and Scowcroft, 1981)

somaclone: Plants derived from any form of cell culture involving the use of somatic plant cells.

somatic: Referring to diploid vegetative or nonsexual parts or body cells.

somatic cell genetics: Asexual genetics involving the study of somatic mutation, assortment, crossing-over, and cell fusion. The cells under study are most often grown in culture.

somatic cell hybrid: A cell plant, or animal that results from the fusion of two animal cells or plant protoplasts derived from somatic cells that differ genetically. (Barski, Sorieul, and Cornefert, 1960)

somatic cell hybridization: A technique that permits hybridization between somatic cells of same or different species in tissue culture.

somatic cell variant: A somatic cell with unique features not shared by the others, as selected for in a screening trial that may follow a mutational event.

somatic crossing-over: Crossing-over at mitosis as opposed to meiosis.

somatic doubling: A disruption of the mitotic process that produces a cell with twice the normal chromosome number.

somatic embryogenesis: In plant culture, the process of embryo initiation and development from vegetative or nongametic cells.

somatic hybrid cell lines: Hybrid cells created by fusing cells from the organism of interest with an immortalized cell line.

somatic hypermutation: The occurrence of a high level of mutation in the variable regions of immunoglobulin genes.

somatic mutation: A mutation in somatic cells. It may lead to a sector or clone of an individual mutant cell that can be recognized in a background of normal cells.

somatic pairing: The phenomenon of close association of homolog in somatic cells, as seen in dipteran giant chromosomes. (Metz, 1916)

somatic segregation: The formation by mitosis of cells differing from one another, either through mutation or somatic crossover in the nucleus or through an unequal assortment of cytoplasmic determinants such as is not required by normal differentiation.

somatoplastic sterility: The collapse of zygotes during embryonic stages due to disturbances in embryo-endosperm relationships.

SOS: The full form of this abbreviation is save our soul, which is, in fact, a DNA repair system controlled by a group of genes. *See* SOS GENES.

SOS box: The region of the promoters of various genes that is recognized by the LexA repressor. Release of repression results in the induction of the SOS response.

SOS genes: In *E. coli,* any of a group of genes that are induced upon DNA damage by physical and chemical genotoxic agents, leading to the inactivation of the LexA repressor. SOS genes are involved in DNA repair, mutagenesis, and cell division, as well as a whole set of plasmid-encoded colicins.

SOS response: Repair systems (RecA, uvr) induced by the presence of single-stranded DNA that usually occurs from postreplicative gaps caused by various types of DNA damage. The RecA protein, stimulated by single-stranded DNA, is involved in the inactivation of the LexA repressor, thereby inducing the response. Also known as *SOS repair system*. (Radman, 1974)

source clone: A source that originates from a single plant or explant within a clone—as from a specific virus tested (SVT) or specific pathogen tested (SPT) individual explant or plant.

source plant: A mother plant or donor plant from which an explant used to initiate a culture is taken.

Southern blotting: Transfer of single-stranded, resistance DNA fragments separated in an agarose gel to a nitrocellulose filter (or other binding matrix), which is then analyzed by hybridization to radioactive or biotinylated single-stranded DNA or RNA probes. The hybrids are detected by autoradiography or a color change, respectively.

spacer DNA: Regions of nontranscribed DNA between transcribed segments. (Brown and Weber, 1968; Birnsteil et al., 1968)

spacer tRNA gene: In *E. coli,* any gene coding for transfer RNA and located in the spacer sequence between the genes coding for 16S and 23S ribosomal RNA or downstream from the 5S rRNA genes. (Lund-Dahberg et al., 1976)

specialized gene: A gene whose activity is under the control of constitutive as well as regulatory factors.

specialized nucleoprotein structures (snurps): These are small ribonucleoprotein particles that help to remove meaningless introns from the message issued by a cell's genes.

specialized transduction: One-way gene transfer in bacteria in which a phage that had been integrated as prophage carries some of its own genome and some of the host's into a recipient cell. Only genes adjacent to the site of integration of the virus in its prophage form can be transferred.

speciation: A process whereby, over time, one species evolves into a different species *(anagenesis* or *phyletic speciation)* or whereby one species diverges to become two or more species *(cladogenesis* or *true speciation).* (Simpson, 1944)

species: (1) A group of actually or potentially interbreeding natural populations that are reproductively isolated from other such groups (Mayrs, 1942); (2) the largest and most conclusive reproductive community of sexual and cross-fertilizing individuals which share a common gene pool (Dobzhansky, 1935); (3) a lineage evolving separately from others and with its own evolutionary role and tendencies (Simpson, 1944). Different species are usually separated from each other by barriers of incompatibility and sterility known as *species barriers.* Spontaneous or experimental hybridization may be difficult or impossible, and if hybrids are obtained, they are generally more or less sterile. *See* INCOMPATIBILITY and STERILITY.

species group: A group of closely related species, usually with partially overlapping ranges.

species selection: Unlike natural selection, which acts at the level of the individual, species selection acts on groups of individuals, or species. Whereas natural selection depends upon an individual's ability to survive, or least put off death, and on its rate of reproduction, species selection depends upon species' ability to survive against extinction and on their rate of speciation.

species swarm: A large number of closely related species. A species swarm can be the result of a population splitting into a number of founder populations that form new species.

specific combining ability: *See* COMBINING ABILITY, SPECIFIC.

specific modifier: A gene that has the specific and perhaps exclusive function of modifying the expression of a gene at another locus.

specific resistance: Host plant resistance to specific biotypes of a pathogen; the interaction of host plant gene-conditioning resistant reaction with a pathogen gene for pathogenicity-conditioning avirulence.

specificity factors: Proteins that reversibly associate with the core component of RNA polymerase and determine which promoters the polymerase will bind to.

specific-locus test: A system for detecting recessive mutations in diploids. Normal individuals treated with a mutagen are mated to testers that are homozygous for the recessive alleles at a number of specific loci; the progeny are then screened for recessive phenotypes. (Russell, 1951)

spheroplast: A bacterial or yeast cell whose wall has been largely or entirely removed. *Compare* PROTOPLAST.

spherosomes: Vesicles in plant cells that appear to perform functions similar to those of lysosomes in animal cells. (Dangeard, 1919; Perner, 1952)

spindle: The microtubule apparatus that controls chromosome movement during mitosis and meiosis.

spindle attachment: That point of a chromatid to which the spindle "fiber" appears to be attached at metaphase and anaphase and that, at anaphase, starts to move toward the pole before the rest of the chromatid. Also referred to as *centromere, insertion region,* or *kinetochore.*

spiral: A coil of the chromosome, chromatid, or chromosome thread at mitosis or meiosis.

spliceosome: It is constituted by five different Usn RNP particles (U1, U2, U4, U5, and U6) during intron splicing. The protein-RNA structure believed to be responsible for splicing. (Brody and Abelson, 1985; Gradbowski, Seiler, and Sharp, 1985)

splicing: The removal of introns from the primary transcript of a discontinuous gene.

splicing junctions: The sequences immediately surrounding the exon-intron boundaries.

splicing, RNA: Joining together separated RNA parts. RNA splicing in eukaryotes entails the removal of introns and the joining of exons of pre-mRNA to produce mature mRNA.

split gene: A discontinuous gene—a gene in which coding sequences are separated by noncoding sequences.

split promoter: The intergenic promoter of eukaryotic genes encoding transfer RNAs that consists of two separated regions within the gene's coding sequence. (Hofstetter, 1981)

spontaneous mutations: Genetic changes produced under normal growth conditions.

spontaneous univalent: This is the spontaneous production of an aberration in which one member of a homologous pair of chromosomes in the cell or an individual is lacking.

spontaneous variation: The variation in plants derived from tissue cultures not exposed to mutagens but occurring as a result of the culture conditions.

sporophyte: In plants, the phase of the life cycle that reproduces asexually by meiospores and is characterized by having the double (usually diploid) chromosome number.

spot test: In mutagenicity testing, an *in vivo* method for detecting mutagens and carcinogens. The method involves placing the agent in the center of a petri dish containing minimal medium that is seeded with auxotrophic bacteria. Revertants from auxotrophy to prototrophy are identified as colonies in a ring around the spot where the potential mutagen was deposited. This test permits characterization of mutagens inducing base-pair substitutions, frameshift mutations, or large deletions. (Ames, 1971)

spreading position effect: Phenomenon in which the effect of an inversion or translocation in inactivating adjacent gene sequences affects more than one such sequence.

ssDNA: Single-stranded DNA.

Sst I: A type II restriction enzyme produced by the bacterium *Streptomyces stanford* that recognizes the DNA sequences shown below and cuts at the sites indicated by the arrows:

```
                   ↓
5′  G  A  G  C  T  C  3′
3′  C  T  C  C  A  G  5′
       ↑
```

stabilizing selection: A type of selection that removes individuals from both ends of a phenotypic distribution divided by deviation of a sample of means.

stable mRNA: An mRNA species that persists in time instead of being rapidly degraded. The stabilization is probably the result of complexing with a protein.

stable RNA: RNA molecules, such as rRNA and tRNA but not mRNA, that are not subject to rapid turnover in the cell.

stage I of clonal multiplication: A step during *in vitro* propagation characterized by the establishment of an aseptic tissue culture of a plant.

stage II of clonal multiplication: A step during *in vitro* plant propagation characterized by the rapid numerical increase of organs or other structures.

stage III of clonal multiplication: A step during *in vitro* plant propagation characterized by preparation of propagule for successful transfer to soil—a process involving rooting of shoot cuttings, hardening of plants, and initiating the change from the heterotrophic to the autotrophic state.

stage IV of clonal multiplication: A step during *in vitro* plant propagation characterized by the establishment in soil of a tissue-culture derived plant, either after undergoing a stage III pretransplant treatment or, in certain species, after the direct transfer of plants from stage I into soil.

staggered cuts: In duplex DNA, these cuts are made when two strands are cleaved at different points near each other.

standard type: The arbitrarily specified genotype used as a basis of comparison for genetic studies. The genes of a standard type are usually symbolized by a "+" sign. All genes deviating from those of standard type are designated by names or symbols. This term is synonymously used with *wild-type*.

starch gel electrophoresis: Zone electrophoresis method in which the support medium is a starch gel. As components of the mixture migrate through the gel under the influence of the electric current, the pore structure of the gel acts as a molecular sieve. As a result, proteins of the same charge-to-mass ratio separate if their molecular sizes are sufficiently different.

start codon: A codon that codes for initiation of protein synthesis, e.g., AUG. Also known as *initiation codon*.

start point (start site): The position on DNA corresponding to the first base incorporated into RNA.

stasipatric speciation: Instantaneous speciation caused by polyploidy.

stasis: The persistence of a species over time without change.

stationary frequency distribution: The representation of a probability density function showing the way the probability in a given situation is distributed over the possible events.

statistic: The actual value of some quantitative character for a sample from which estimates of parameters may be made.

statistical distribution: The classification of a population under study on the basis of a statistical parameter into its alternate phenotypes.

statistical genetics: This branch of genetics deals with the study of theoretical population genetics that is predominantly statistical and mathematical in nature. *See* BIOMETRICAL GENETICS. (Wright, 1942)

statistical proteins: In absence of a genetic code, proteins synthesized will have a random sequence of amino acids.

stearic fit model: An early hypothesis to explain the relationship between an amino acid and its codon on the basis of the stearic relationship between the two.

stemline: The genetically uniform cell type that characterizes and mainly propagates a normal tissue, a tumor, or a tissue-culture cell line. Besides the stemline, other cell types may be present in the cell populations mentioned as secondary stemlines or more temporary lines, potentially capable of developing into new stemlines. The stemline concept, being of mainly theoretical significance, is for practical purposes often replaced by the stemline chromosome number (the numerical stemline) or s, which is the modal number of chromosomal elements without regard to frequency or structure of chromosome types. (Levan and Hauschka, 1953)

stem-loop structure: A lollipop-shaped structure formed when a single-stranded nucleic acid molecule loops back on itself to form a complimentary double helix (stem) topped by a loop.

step allelomorphism. The occurrence of a series of multiple allelomorphs with overlapping effects that can supposedly be related to a linear order in the distribution of units of change within the gene.

stepladder gel: A DNA-sequencing gel. The numerous bands in each lane give the appearance of a stepladder.

sterile: (1) Without life; (2) inability of an organism to produce functional gametes.

sterility: The failure to complete fertilization and obtain seed as a result of defective pollen or ovules, or other aberrations.

stickiness: A sort of chromosomal "agglutination" of unknown nature that results in a pycnotic or sticky appearance of chromosomes which may give rise to sticky adhesions between two or more chromosomes and to the formation of "sticky bridges" at anaphase. (Beadle, 1932)

sticky association: Any association of two or more chromosomes at mitosis and meiosis due to nonspecific sticky adhesions rather than to homologous chromosomes. (Price, 1956)

sticky end: An end of a double-stranded DNA molecule where there is a single-stranded extension. Sticky ends may be extended at

5′ end (e.g., those produced by EcoRI restriction endonuclease) or at 3′ end (e.g., those produced by HsaI restriction endonuclease).

stigma: The portion of the pistil that receives the pollen.

stimulon: A set of genes that respond to a single stimulus (regulon, operon). Stimulons may overlap if they contain an operon with multiple promoters that have different regulatory properties.

stochastic: A process with an indeterminate or random element as opposed to a deterministic process that has no random element.

stock: An artificial mating group.

stock plant: The source plant from which cuttings or explants are made. These are usually maintained carefully in an optimum state for (sometimes prolonged) explant use. Preferably they are certified, pathogen-free plants.

stop codon: The codons that provide the signal for termination of polypeptide chains, e.g., UAA, UAG, and UGA.

strain: A group of individuals from a common origin. Generally, a more narrowly defined group than a variety.

strain building: The improvement of cross-fertilizing plants by any one of a number of methods of selection.

strand: (1) A visible linear structure (chromatid) in the chromosome; (2) one of the two polynucleotides in the double helix.

strand displacement: It is a mode of replication of some viruses in which a new DNA strand grows by displacing the previous (homologous) strand of the duplex.

strand transfer repair: A DNA repair mechanism in which a recombination-like process forms gaps opposite the induced lesion in a replicating template strand. These gaps are then filled in via strand transfer from the previously replicated normal strand.

strand transferase: Any of the enzymes that promote homologous pairing by forming a complex between a double-stranded DNA molecule and a homologous single-stranded DNA molecule (Shibata et al., 1979)

stringent factor: A protein that catalyzes the formation of an unusual nucleotide (guanosine tetraphosphate) during the stringent response under amino acid starvation.

stringent plasmid: A plasmid whose copy number is under strict control and therefore has only one or two copies per chromosome, i.e., a low copy-number plasmid, such as the F plasmid.

stringent replication: The limitation of single-copy plasmids to replication with the bacterial chromosome.

stringent response: A translational control mechanism of prokaryotes that represses tRNA and rRNA synthesis during amino acid starvation.

strong promoter: A promoter for which RNA polymerase has a high affinity and which directs the synthesis of large amounts of mRNA.

structural alleles: Alleles determined on the basis of a recombination test.

structural change: Change in the genetic structure of one or more chromosomes. May be intraradial or extraradial with respect to arms; internal, fraternal, or external with regard to chromosomes; and eccentric or dicentric with respect to the possession of a centromere or to the direction of a segment in relation to the centromere. (Darlington, 1929a; 1937b)

structural gene: A gene that codes for an RNA molecule or protein other than a regulatory protein. (Jacob and Monod, 1959)

structural hybridity: Heterozygosity for a chromosomal rearrangement.

subclone: A method in which smaller DNA fragments are cloned from a large insert which has already been cloned in a vector.

subculture: With plant cultures, this is the process by which the tissue or explant is first subdivided, then transferred into fresh culture medium. *See* PASSAGE.

subinert chromosome: A chromosome with a slight genetic effect is said to be subinert. *See* INERT GENE.

sublethal mutation: *See* VITALITY MUTATION.

submetacentric chromosome: A chromosome where the centromere is nearer to one end than the other, resulting in the arms not being of equal length.

subspecies: One of the several geographic or similar subdivisions between which interbreeding takes place at a reduced level. (Wettstein, 1898; Du Rietz, 1930)

substitution haploid: A haploid individual having one chromosome replaced by that of a related species. ($2x - 1 + 1$, $3x - 1 + 1$). These are derived from alien substitution lines.

substitution line: A chromosome transferred into a recipient variety through monosomic analysis techniques.

substitutional load: The cost to a population of replacing an allele with another in the course of evolutionary change.

substrain: A substrain can be derived from a strain by isolating a single cell or groups of cells having properties or markers not shared by all cells of the parent strain.

substrate race: A local race selected to agree in its coloration with that of the substrate, for example, a black race on a lava flow.

substrate subspecies: A population whose coloration agrees with the soil or rocky substrate on which it lives, such as a pale-colored population living on a sandy desert.

subtelocentric chromosome: A chromosome whose centromere lies between its middle and its end but closer to the middle.

subtraction hybridization: A method of screening cDNA libraries whereby two probes, differing in one or just a few sequences, are hybridized together so that the sequences common to the two populations anneal. The nonannealed sequences are then used to prepare a cDNA library and/or a probe for screening.

subvital gene: A gene that causes the death of some proportion, but not all, of the individuals that express it.

super dominance: *See* TRUE OVERDOMINANCE.

super multigene family: Genes that have considerable homology but do not perform the same function.

super repressed: Uninducible.

super suppressor: A mutation that can suppress a variety of other mutations; typically a nonsense suppressor. (Hawthrone, and Mortimer, 1963; Gilmore, 1967)

supercoiling: Negative or positive coiling of double-stranded DNA that differs from the relaxed state.

supergene: The advantageous grouping of genes within an inversion; the entire gene complex is inherited as a whole. (Darlington and Mathar, 1949)

superlethal mutation: *See* VITALITY MUTATION.

supernumerary chromosomes: Chromosomes present, often in varying numbers, in addition to the characteristic, relatively invariable complement. *See* B CHROMOSOMES. (Longley, 1937)

superspecies: A monophyletic group of entirely or essentially allopatric species that are too distinct to be included in a single species.

supplementary genes: Two dominant genes that when present together produce a novel phenotype. These genes yield a 9:3:4 ratio in F_2 progeny.

suppression: A mutation in a gene that reverses the effect of a previous mutation in a different gene. This term also refers to the rescue of an allele by means of another gene. (Sturtevant, 1920)

suppressor, extragenic: Usually a gene coding a mutant tRNA that reads the mutated codon either in the sense of the original codon or gives an acceptable substitute for the original meaning.

suppressor gene: A gene that, when mutated, apparently restores the wild-type phenotype to a mutant of another locus.

suppressor, intragenic: A compensating mutation that restores the original reading frame after a frameshift.

suppressor mutation: A mutation at a site different from the original mutation that can restore a lost function.

suppressor rescue: A transferred gene is tagged with an amber suppressor gene to isolate a human oncogene.

suppressor tRNA: Any transfer RNA species that, due to mutation in a transfer RNA gene, shows altered coding specificity and is able to translate nonsense or missense codons. Nonsense and missense suppressor tRNA is frequently modified in the anticodon loop or in an adjacent region. Frameshift suppressor tRNA compensates for frameshift mutations by reading nontriplet codons. (Capecchi and Gussin, 1965)

suppressor-sensitive mutation: A mutation whose phenotype is suppressed in a genotype that also carries an intergenic suppressor of that mutation, e.g., amber, ocher, or opal.

supraoperon control: The coordinate control of a gene cluster with multiple operator and promoter sites that consists, in essence, of several interacting operons controlling different biochemical pathways. (Roth and Nester, 1971)

surrogate genetics: A branch of genetics that deals with the introduction of a manipulated DNA into a living nucleus where its expression can be monitored. Effects of directed DNA changes, changes in RNA transcript, and changes in protein encoded by it can all be studied. *See* REVERSE GENETICS.

survival of the fittest: In Darwin's evolutionary theory, survival of only those individual organisms of a generation that are best able to obtain and utilize resources (fittest). This phenomenon is the cornerstone of Darwin's theory.

susceptible: A characteristic of a host plant such that it is incapable of suppressing or retarding an injurious pathogen or other factor.

suspension culture: A type of culture in which cells, or aggregates of cells, multiply while suspended in liquid medium.

Svedberg unit (S): *See* SEDIMENTATION COEFFICIENT.

switch gene: A gene that causes the epigenotype to switch to a different developmental pathway. *See* OLIGOGENES.

swivel: The place in DNA at which replication starts. (Cairns, 1963)

swivelase: *See* DNA GYRASE.

symbiosis theory of origin of eukaryotic cells: A prokaryotic cell having a flexible-cell, membrane ingested, bacterial cell that evolved into mitochondria. Spirochetes become ingested and converted into nuclear fibers, centrioles, and flagella of a primitive unicellular eukaryote. Blue-green algae cells become incorporated into colorless, unicellular eukaryotes to form a simple, green flagellate. This is the most favored theory. (Margulis, 1970)

sympatric hybridization: This is hybridization between individuals of populations within the range and habitat of the parent species. This type of hybridization may be common in parasites.

sympatric speciation: Speciation in which the evolution of reproductive isolating mechanisms occurs within the range and habitat of the parent species.

Syn 0, Syn 1, Syn 2: Symbols for designating the original population, first synthetic generation (progeny of Syn 0), and second synthetic generation (progeny of Syn 1).

synapsis: The pairing of homologous chromosomes that occurs in prophase I of meiosis.

synaptomere: Any of the specific points where the synaptonemal complex formation starts. (King, 1970)

synaptonemal complex: A structure that helps in the pairing of homologous chromosomes during the pachytene stage, consisting of a tripartite ribbon of parallel, dense, and lateral elements surrounding a median complex. (Moses, 1958)

synchronous culture: A culture in which the cell cycles (or a specific phase of the cycle) of a proportion of the cells (often a majority) are synchronous.

syncytium: A single cell with many nuclei.

syndrome: A group of symptoms that occur together and represent a particular disease.

synergid: In higher plants, any of the (two) haploid nuclei found with the egg at the micropylar end of the embryo sac.

synergistic gene expression: When two transcription activation factors, together, lead to a higher rate of transcription than either of the factors does individually.

syngamy: The union of the nuclei of sex cells (gametes) in reproduction.

synkaryon: The hybrid cell that results from the fusion of the two nuclei it carries.

synonymous codons: Triplet codons that code for the same amino acid. For example, UCU, UCC, UCG, UCA, AGU, and AGC all code for serine and are therefore synonyms.

synonymous mutation: A class of gene mutations that represent base pair substitutions in DNA that do not result in the substitution of a different amino acid into the encoded protein. This is due to the degeneracy of the genetic code. (Sonneborn, 1965a)

syntenic: Genetic loci that lie on the same chromosome. (Renwick, 1971)

syntenic genes: Genes that are believed to be linked (i.e., reside on the same chromosome) because of their concordant association along with a specific marker gene known to be located on a particular chromosome.

synteny: The occurrence of two or more molecular markers or genetic loci on the same chromosome. Depending on intergene distance(s) involved, they may or may not exhibit nonrandom assortment at meiosis.

synteny test: A test that determines whether two loci belong to the same group by observing concordance in hybrid cell lines.

synthetic lethal: Lethal chromosomes derived from normally viable chromosomes by recombination (as a result of crossing-over).

synthetic phase: A stage in interphase during which chromosome replication occurs. At the end of this phase the amount of DNA doubles and each chromosome has two chromatids.

synthetic plasmid: A circular DNA molecule that is constructed from a replicon and one or more gene DNA sequences. The synthetic plasmids have a genetic makeup that differs from wild-type plasmids. They are used as cloning vehicles.

synthetic polyribonucleotides: RNA made *in vitro* without a nucleic acid template, either by enzymatic or chemical synthesis.

synthetic populations: Populations derived from inter se crossing in diallel fashion of a number (generally 4 to 8) of inbreds of known high per se performance and combining ability and subsequently maintained by open pollination. Composites are also synthetic populations but involve any number and any kind of parental components—inbred, variety, single crosses, double crosses, etc.—of unknown performance.

synthetic theory of evolution: The current extension of Darwin's theory of evolution in which mutation, recombination, natural selection, geographic isolation, and reproductive isolation play major roles.

synthetic variety: A variety that originates as a known mixture of cross-fertilizing genotypes—advanced generations of open-pollinated seed mixtures of a group of strains, clones, inbreeds, or hybrids among them. The component units are propagated and the synthetics reconstituted at regular intervals.

synthetics: Populations comprising of seed mixture of genetically known strains, clones, inbreds, or hybrids among them, maintained by open-pollination for a specified number of generations. The component units are propagated and the synthetics reconstituted at regular intervals.

systematic mutation: Mutations of major effect presumed to give rise to new species or higher categories at a single step.

systematic pressure: One of the nonrandom evolutionary pressures—selection, mutation, or migration.

T

t: (1) The ratio of an observed deviation to its estimated standard deviation; (2) indicates a translocation in cytogenetics.

T I, T II: Symbols for telophase of first (T I) and second meiotic division (T II)

T phages: T phages (T1-T7) are virulent phages that contain double-stranded DNA and affect its antigenic properties.

tactaac box: A consensus sequence surrounding the lariat branch point of eukaryotic mRNA.

tailing: This is nontranscriptive addition of homopolymeric nucleotide to the 3′ end of a restriction fragment by use of enzyme terminal transferase.

tandem duplication: Situation in which a segment of a chromosome is duplicated and the extra copy or copies inserted in line in the chromosome. Thus, the genes for large ribosomal RNA are tandemly duplicated in eukaryotes.

tandem genes: Genes that are arranged tandemly in the chromosomes, especially multiple-copy genes where the copies are tandemly duplicated. *See* TANDEM DUPLICATION.

tandem repeats (array): Direct repeats that are adjacent to each other with no intervening DNA.

tandem satellites: Two short segments of chromosomes separated from each other by a constriction and from the main body of the chromosome by a second constriction. (Taylor, 1926)

tandem selection: Exercising selection for a single trait at a time in a population. In contrast, more than one character is selected for simultaneously under the index-selection scheme.

tandon: A multiple-copy, tandomly repeated gene. (Tartof, 1975)

target nucleic acid: A nucleic molecule that is to be detected in a hybridization reaction or isolated by molecular cloning.

target site: Restriction endonucleases bind to DNA at a specific sequence known as their recognition site and then cut the DNA within a sequence known as the target site. For type II restriction endonucleases, the kind most commonly used in genetic engineering experiments, the recognition site and the target site are one and the same thing.

target site duplication: A sequence of DNA that is duplicated when a transposable element inserts, usually found at each end of the insertion.

target theory: A theory that predicts response curves based on the number of events required to cause the mutational phenomenon in a gene. This theory was used to estimate the size of the gene.

targeted gene replacement: One of the most powerful methods of discovering what a gene does. A gene is knocked out and its effect is observed on the organism. In this way, researchers can create an organism bearing chosen mutations in any known gene. (Capecchi, 1994)

targeted gene transfer: The transfer of gene sequences directed to a specific site in the genome. The vector used to introduce the new gene into the cells must carry nucleotide sequences identical to those of the DNA at the genomic site where the gene should integrate by homologous recombination. Targeted transfer can be used to correct or inactivate genes.

targeted mutagenesis: Use of a mutagen, such as ultraviolet light or nitrogen mustard, that is specific for a particular target within the cellular DNA. (Witkin and Wermudsen, 1978)

TATA binding protein (TBP): TBP is essential for all transcription. Its detection in eukaryotic cell nuclei and *Archaebacteria* shows that transcription mechanisms of *Archaebacteria* and eukaryotes are fundamentally homologous. (Rowlends, Baumann, and Jackson, 1994)

TATA box: A nucleotide sequence at the 5′ end of eukaryotic messenger RNA encoding genes (about 30 bp long and upstream from the genome location of the cap site) that has a critical role in the initiation of genetic transcription. TATA boxes are general promotor elements that are recognized by a common transcription factor (Hogness-Goldberg box). The TATA box is analogous to the Pribnow box of prokaryotic promoters found around position-30. (Goldberg, 1979)

tautomer: An isomer of an organic substance, differing from other forms of the substance in the position of a hydrogen atom and a double bond.

tautomeric shift: A reversible shift of proton position in a molecule. Based on nucleic acids shift between keto and enol forms or between amino and imino forms.

tautomerization: The process of shifting of hydrogen atoms from one position to another in a purine or in a pyrimidine base. The product of this process is known as *tautomer*.

T-DNA: Complete Ti plasmid is not found in plant tumor cells, only 20 kb DNA of Ti plasmid called T-DNA is found integrated in plant nuclear DNA. Genes of T-DNA are eukaryotic in origin and have been captured by Ti plasmid during evolution.

TEL: A DNA segment containing telomeric sequences. TEL DNA is required for replication and maintenance of linear chromosomes.

These are highly conserved among widely divergent species. TEL sequence in yeast is $(C_{1-3}A)_n$, in Tetrahymena $(C_4A_2)_n$, and in mammals, birds, and reptiles $(C_3TA_2)_n$.

telocentric chromosome: A chromosome with a terminal centromere. (Darlington, 1939a)

teloisodisomic: A cell or individual in which one chromosome pair is missing but a telocentric for one arm of the missing pair and an isochromosome for the same arm are present. (Kimber and Sears, 1968)

teloisotrisomic: An individual deficient in one chromosome but which has a telocentric chromosome and an isochromosome for the same arm of the missing chromosome.

telomerase: The enzyme that maintains the ends of eukaryotic chromosomes by synthesizing telomeric repeat sequences. It is composed of RNAs and protein. (Szostak and Blackburn, 1982)

telomere: The end of a chromosome that provides stability to the chromosome, prevents fusion with other natural or broken ends, and allows replication without loss. It consists of simple DNA repeat sequences (2 to 10 bp long) fitting the formula $5'\ C_n(A/T)_m/G_n(T/A)_m\ 3'$ ($n = 1$ to 8; $m = 1$ to 4) that defines orientation (C/A-rich strand running $5'$ to $3'$ from the end toward the interior of the DNA molecule). The C/A-rich strand usually contains nicks, and the G/T-rich strand extends into a unique terminal structure. Telomeres have non-ligatable single-strand gaps, and extreme ends are blocked by a hairpin loop at the $3'$ end in one strand, have nonnucleosomal protein complexes, and subtelomeric middle-repetitive sequences are present, unlike the internal chromosome regions. (Muller, 1940a)

telomere fusion: The end-to-end joining of two chromosomes to form one pseudodicentric chromosome by terminal rearrangement. If one centromere of the pseudodicentric is suppressed, telomere fusion may result in a stable chromosome. (Niebuhr, 1972)

telomere terminal transference: A ribonucleoprotein enzyme that can synthesize the tandem repeat units of telomere to the $3'$ end of telomeric primers without template. The RNA component of the enzyme contains the complementary sequence of the telomeric repeats that it synthesizes.

telophase: The concluding stage of nuclear division, characterized by the reorganization of interphase nuclei. (Heidenhain, 1894)

telophase I: The concluding stage of meiosis I that is characterized by the reappearance of the nuclei which carry a haploid (2n) number of chromosomes.

telophase II: The concluding stage of meiosis II that is characterized by the reappearance of the nuclei, four in number, which carry haploid (2n) number of chromosomes.

telosomic trisomy: The extra chromosome is a telocentric fragment chromosome homologous to one arm of a chromosome pair in the standard complement. (Burnham, 1962)

telotertiary compensating trisomic: A compensating trisomic in which the missing chromosome is compensated for by one telocentric chromosome and one tertiary chromosome.

telotrisomic: A trisomic in which the extra chromosome is a telocentric chromosome for one arm. Also called *monotelotrisomic*. (Kimber and Sears, 1968)

teminism: *See* REVERSE TRANSCRIPTION.

temperate phage: A phage (virus) that invades but may not destroy (lyse) the host (bacterial cell). However, it may continue into the lytic cycle. *Compare* VIRULENT PHAGE. (Jacob et al., 1953)

temperature (symbiotic) phage: A bacterial virus that tends to take up residence in the host cell without destroying it.

temperature-sensitive mutation: An organism with an allele that is normal at a permissive temperature but mutant at a restrictive temperature.

template strand: The polynucleotide of a gene that acts as the template for RNA synthesis during transcription.

temporal gene: Any gene that determines the relative expression of structural genes in a particular organ or at a particular developmental time. Temporal genes are divided into two classes: class I temporal genes are trans-acting genes that lie at a distance from the structural genes whose expression they influence; class II temporal genes are *cis*-acting genes that lie in close proximity to the structural gene. (Paigen and Ganschow, 1965)

terminal affinity: The property by which chromosomes are held together end to end from diplotene until first metaphase of meiosis (by a terminal chiasma) or are brought together in this way at metaphase. (Darlington, 1932a)

terminal CCA: At the 3′ end of all tRNA there is a sequence, CCA, to the adenosine moiety to which is attached the amino acid.

terminal chiasmata: The association of homologous chromosomes at meiosis, involving only the chromosome ends. This may be the result of a crossing-over event at the chromosomal extremity, or of terminalization of chiasmata, which earlier were positioned some distance away from the chromosome terminus.

terminal deletion: A deletion involving the end of a chromosome.

terminal deoxynucleotidyl transferase (TdT): A unique DNA polymerase that catalyzes the polymerization of 5′-dNTPs onto the 3′-OH terminus of an oligodeoxynucleotide. TdT does not require a template but has an absolute requirement for an initiator molecule $(dNMP)_n$; where n = at least 3. (Bollum, 1960)

terminal gene: A gene situated at the extreme end of the chromosome telomere.

terminal redundancy: The repetition of the same sequence at both ends of (for example) a phage genome. (Streisinger, Edger, and Denhardt, 1964)

terminal riboadenylate transferase (TrT): An enzyme that catalyzes the transfer of adenylate residues from ATP to the 3′-OH group of certain polyribonucleotides in the presence of Mn^{2+}. TrT may be involved in the processing of heterogeneous nuclear RNA to messenger RNA in eukaryotic cells.

terminalization: The movement of the chiasma toward the ends of the bivalent during meiosis. The number of chiasma is thus reduced from diplotene to metaphase I.

termination codon: One of the three codons (5′-UAA-3′, 5′-UAG-3′, and 5′-UGA-3′ in the standard genetic code) that mark the position where translation of an mRNA should stop.

termination factor (TF): The protein required to obtain release of a newly synthesized polypeptide chain from tRNA. (Ganoza, 1966)

terminator codon: *See* NONSENSE CODON.

terminator sequence: A sequence in DNA that signals the termination of transcription to RNA polymerase.

terminator stem: A configuration of the leader transcript that signals transcription termination in attenuator-controlled amino acid operons.

tertiary monosomic: An individual in which two arms of two nonhomologous chromosomes are missing but the remaining two arms are joined to constitute a tertiary chromosome.

tertiary protein structure: The formation of disulfide bridges between cysteines as well as the further folding of a protein beyond the secondary structure.

tertiary trisomic: A trisomic in which the additional chromosome is a translocated one, i.e., consists of two arms of two nonhomologous chromosomes. (Belling and Blakeslee, 1926)

test cross: A genetic cross between an individual (generally of dominant phenotype) and the one having a recessive phenotype. It is used to determine the genotype (homozygous or heterozygous) of an individual of dominant phenotype. It is also used to construct linkage maps. (Bridges, 1934)

test of significance: A test designed to assess significance to distinguish differences due to sampling errors from differences indicating discrepancies between observation and hypothesis.

test systems: The organisms and methodology used to test mutagenicity and/or carcinogenicity of environmental, chemical, and physical agents.

tester: An individual homozygous for one or more recessive alleles, often used in a test cross.

tetracyclines: Group of broad spectrum antibiotics obtained from various species of the mold streptomyces. They bind to the 30S subunit of the bacterial ribosome, preventing attachment of the amino-acyl tRNA.)

tetrad: The four haploid cells that result from a single meiotic event in a eukaryotic germ cell. (Němec, 1910)

tetrad analysis: The genetic analysis of the products of a single meiosis division.

tetrad divisions: The first and second meiotic divisions which the spore mother cell undergoes to form four spores.

tetramer: Structure resulting from association of four subunits.

tetranucleotide hypothesis: The hypothesis, based on incorrect information, that DNA could not be the genetic material because its structure was too simple—that is, that repeating subunits contain one copy each of the four DNA nucleotides.

tetraploid: A polyploid cell, tissue, or organism with four sets of chromosomes (4n).

tetrasomic: Having two extra chromosomes of a given kind, making four of the kind in question (2n + 2). (Blakeslee, 1921)

tetrasomy: The condition in which the cells of an organism contain one chromosome of the basic complement represented four times.

tetratype: In fungi, a tetrad of spores that contains four different types; for example, AB, aB, Ab, and ab. Crossing-over has occurred in such a tetrad.

tetravalent: An association at first meiotic division of four chromosomes that are completely or partially homologous. *See* QUADRIVALENT.

theory of inheritance of acquired traits: Acquired characters are inherited and adaptive and thus, account for evolutionary change. *See* LAMARCKISM.

thermoregulation of a gene: Regulation of gene expression is controlled by temperature.

thin layer chromatography (TLC): A chromatographic method in which a glass plate is covered with a thin layer of inert, absorbent material (e.g., cellulose or silica gel), and the materials to be analyzed are spotted near the lower edge of the plate. The base of the plate is then placed in a solvent, which rises up the plate capillary action, separating the constituents of the mixtures. The principles involved are similar to those of paper chromatography and, like paper chromatography, two-dimensional methods can also be employed.

thin layer gel filtration: A technique in which molecules are separated on a flat plate (typically 20 × 20 cm) on which a layer of xerogel 0.5 × 1.0 mm thick is spread.

thirty (30) nm chromatin fiber: The substructure of chromatin that consists of a possible helix array of nucleosomes in a fiber approximately 30 nm in diameter.

Thomas circle: A double-stranded DNA fragment from eukaryotic chromosomes that can form circular structures after treatment with exonucleases. The formation of Thomas circles is assumed to be due to the presence of repetitious DNA sequences in the fragments. (Schachat and Hogness, 1973)

three prime (3′)-OH terminus: The end of a polynucleotide that terminates with a hydroxyl group attached to the 3′ carbon of the sugar.

three-breed crossing: A system of breeding in which breed A is crossed with B and the F_1 with C. This hybrid is then crossed with B and the product is crossed with A to produce a hybrid which is crossed with C. The next cross is with B and then A and so on.

three-point test cross: A trihybrid test cross (e.g., ABC/abc x abc/abc, etc.) used primarily in chromosome mapping.

three-way cross: The progeny of a cross between a single-cross and an inbred-line or pure-line variety.

threshold: A term used in studying the effect of radiation. Below a certain dose, if there is a threshold, there is no measurable effect. No threshold exists for most genetic effects.

threshold character: A term used for those phenotypic characters whose segregating distributions are phenotypically discontinuous but whose inheritance is multigenic like that of quantitative characters. Threshold characters may segregate into many discontinuous phenotypic classes. (Dempster and Lerner, 1950)

threshold model: A model proposing that there is a variation of individual liability for a particular trait; the trait appears only when this liability exceeds a threshold value.

thymidine: The deoxyribonucleoside that contains the pyrimidine thymine.

thymidylic acid: The deoxyribonucleotide that contains the pyrimidine thymine.

thymine: 5-methyl uracil, also written as 2, 6-dioxy, 5-methyl pyrimidine. One of the nitrogenous bases found in DNA.

thymine dimer: Hydrogen bonding of two molecules of thymine by action of ultraviolet light.

Ti plasmid: The large plasmid found in *Agrobacterium tumefaciens* cells and used as the basis for a series of cloning vectors for higher plants. (Zaenen et al., 1974)

tier system: In mutagenicity testing, a test system in which potentially mutagenic chemicals detected in a relatively simple test are subjected subsequently to more biologically complex tests until they are either eliminated from further testing or bioassayed in full-scale animal systems.

tissue culture: The growth of cells, tissues, or organs in suitable media *in vitro*. Such media must normally be sterile, correctly pH balanced, and contain all the necessary micro and macronutrients, carbohydrates, vitamins, and hormones for growth. Studies of such cultures have shed light on physiological processes that would be difficult to follow in the living organism. The cytokinins were discovered through work on tobacco pith tissue culture.

Tn 3: A bacterial transposon that carries the gene for ampicillin resistance as well as genes for its own transposition. Tn 3 was the first transposon to have its DNA sequence completely determined.

Tn 5: A bacterial transposon that carries a gene encoding resistance to the antibiotics kanamycin and neomycin, as well as genes involved in transposition.

Tn 10: A bacterial transposon that carries the gene for tetracycline resistance as well as genes involved in its own transposition.

tolerable mutations: The mutations that affect function of the gene to such a degree that they do not have a drastic effect on the individual and are tolerable. Such mutations accompany the process of evolution and play an important role in speciation. Tolerable mutations are of two types—neutral and favorable.

tolerance: The ability of plants to survive in the presence of destructive pathogens, insects, or environmental conditions.

top cross: The cross of a line to be evaluated with an open-pollinate tester; the F_1 produced by crossing an inbred line and a commercial variety; an outcross of selections, clones, lines, or inbreds to a common pollen parent. In maize, commonly an inbred-variety cross.

topcross progeny: Progeny from out-crossed seed of selections, clones, or lines to a common pollen parent.

topoisomerases: An enzyme that can relieve (or create) supercoiling in DNA by creating transitory breaks in one (type I) or both (type II) strands of the helical backbone. *See* DNA TOPOISOMERASES. (Wang and Liu, 1979)

topological isomers (topoisomers): Forms of DNA with the same sequence but differing in their linkage number (coiling).

totipotency: The property of a cell (or cells) whereby it retains the potential of developing into a complete and differentiated adult organism.

totipotent cell (or nucleus): The state of an undifferentiated cell (or nucleus) that, when isolated, can give rise to any and all adult cell types, as compared with a differentiated cell whose fate is determined.

tra genes: The genes involved in the transfer of a plasmid from one cell to another.

tracer: A radioactivity-labeled nucleic acid component included in a reassociation reaction in amounts too small to influence the progress of reaction.

trailer segment: The untranslated segment that lies downstream of the nonsense codon of an mRNA molecule.

trait: An attribute that varies from one individual to another. *See* BIOLOGICAL PROPERTY.

***trans*-acting:** Referring to mutations of, for example, a repressor gene, that act through a diffusible protein product; the normal mode of action of most recessive mutations.

***trans*-arrangement:** The linkage of the dominant allele of one pair and the recessive of another on the same chromosome.

***trans*-configuration:** A double heterozygote where each of the dominant alleles are in the opposite homolog. (Haldane, 1942)

transcribed spacer: Part of an rRNA transcription unit that is transcribed but discarded during maturation; that is, it does not give rise to part of rRNA.

transcript: An RNA copy of a gene.

transcript analysis: Experiments designed to determine which regions of a DNA molecule are transcribed into RNA.

transcription: The synthesis of a stand of RNA using a single strand of a DNA molecule as a template (Hayashi, Hayashi, and Spielgelman, 1964; Szybalski et al., 1970)

transcription bubble: A region of locally melted DNA in the process of being transcribed by RNA polymerase.

transcription factors: Eukaryotic proteins that aid RNA polymerase to recognize promoters. Analogous to procaryotic sigma factors.

transcription pausing: The phenomenon wherein both prokaryotic and eukaryotic RNA polymerases show slow genetic transcription at transcription pause sites when particular regions of a DNA molecule are being transcribed, i.e., chain elongation does not occur at a constant rate. Pausing may be necessary for proper regulation of transcription and occurs immediately after transcription of regions with dyad symmetry by formation of a hairpin near the $3'$ end of a growing transcript.

transcription termination: The cessation of transcription caused by a specific terminator structure.

transcription unit: The distance between sites of initiation and termination by RNA polymerase; may include more than one gene.

transcription unit design: Refers to whether one or more mRNAs encoding different proteins are produced from a primary transcript. Two types of transcriptional units (simple and complex) are known.

transcriptional activator: Any DNA-binding protein that stimulates transcription by interacting with DNA sequence motifs (transcription control regions), RNA polymerase, or both. Some transcriptional activators have two surfaces: (a) The DNA binding region that positions the protein on DNA; and (b) the activating region that interacts with a target protein to promote initiation of transcription. (Brent and Ptashne, 1985)

transcriptional antitermination: A mechanism that allows transcription termination signals of an operon to be overcome.

transcriptional control: Control of gene expression at the level of genetic transcription that can be exerted through regulation of transcription initiation or transcription termination through selective posttranscriptional processing. Transcriptional control may be negative, in which transcription is turned off by repressor molecules that can bind to an operator and prevent RNA polymerase from traveling across the operator, or positive, in which transcription is turned on by a positive controller which may or may not be a protein and does not involve release of repression.

transcriptional enhancer: Any of a class of *cis*-acting DNA sequences (usually 200 bp long) that are able to activate the transcription by RNA polymerase II of cellular and viral genes in a manner relatively independent of position and orientation. Transcription activation can occur over distances greater than 6 kb from either the 5′ or 3′ end of a gene. Many enhancers include an octamer sequence motif ATTTGCATT. Frequently, enhancers are found near Z-DNA and are localized in introns. *See* ENHANCERS. (Gruss, Dhar, and Khoury, 1981)

***trans*-dominant mutation:** A mutation in a regulatory gene that can effect the expression of a structural gene on another chromosome. *See* *CIS*-DOMINANT MUTATION.

transduced element: A fragment of DNA transferred from one cell to another in the process of transduction.

transductant: A cell that has been transduced.

transduction: The nonsexual transfer of desired genetic information coded on a DNA segment from one organism to the other through an appropriate molecular (DNA) vector (such as bacteriophage/virus-mediated or plasmid-mediated transfer) process of genetic engineering. (Zinder and Lederberg, 1952)

transfection: (1) The introduction of foreign DNA into eukaryotic cells; (2) the transfer, for the purposes of genomic integration, of naked, foreign DNA into cells in culture. The traditional microbiological usage of this term implied that the DNA being transferred was derived from a virus. The definition as stated here is the one in use to describe the general transfer of DNA irrespective of its source. *See* TRANSFORMATION. (Földes and Trautner, 1964)

transfer reaction: A reaction that attaches an amino acid with its specific tRNA. This reaction is conducted by a specific aminoacyl synthetase.

transfer RNA (tRNA): Abbreviated as tRNA. One of the small RNA molecules that acts as an adapter during translation and is responsible for decoding the genetic code. (Hoagland, 1959)

transfer RNA identity: Correct recognition of transfer RNA by aminoacyl-tRNA synthetases is central to the maintenance of translational fidelity. The hypothesis that synthetases recognize anticodon nucleotides was not accepted until relatively recently. Implementation of new methodology has clearly shown that anticodon is indeed important for 7 of the 20 *E. coli* isoaccepting groups. For many of the isoaccepting groups, the acceptor stem or position 73 (or both) is important as well. (Margaret, Jeffrey, and Abelson, 1994)

transformant: A cell that has been transformed by the uptake of naked DNA.

transformation: (1) The conversion of normal eukaryotic cells to malignant cells with the ability for continuous growth in culture; (2) incorporation of genetic material into prokaryotic cells; (3) in plant cell culture, the introduction and stable genomic integration of foreign DNA into a plant cell by any means, resulting in a genetic modification; or (4) nonsexual transfer of desired genetic information coded on a DNA segment (naked) from one organism to the other, directly through incubation or through bioengineering.

transformation, eukaryotic: Conversion of eukaryotic cells to a state of unrestrained growth in culture, resembling or identical with the tumorgenic condition.

transformed cells: Cells that have become capable of sustained proliferation *in vitro*.

transforming principle: The chemical substance responsible for transforming Streptococcus bacteria from one serotype to another, shown by Avery, MacLeod, and McCarty (1944) to be DNA.

transgenation: A heritable change taking place at a single gene locus, presumably caused by chemical alteration of the individual gene—a mutation proper, point mutation, or genovariation.

transgenic: An organism containing genetic material artificially placed there from another species. (Gordon and Ruddle, 1981)

transgenome: A genetic element transferred by chromosome-mediated gene transfer. (Athwal and McBridge, 1977)

transgenosis: Genetic transformation in very evolutionally distant organisms. *See* TRANSFORMATION.

transgression: The origin in the F_2 or later generations of segregation products that are more extreme than either of the parents.

transgressive: It is a phenomenon in which some individual in the progeny shows characters exceeding in value those shown by either of the parents. The transgressive inheritance is usually seen in crosses involving quantitative characters.

transgressive breeding: Isolation of gene combinations (recombinants) possessing new characters or a new intensity of traits surpassing both the parents in F_2 onward. Development of productive variety through "assembling productivity alleles" is an example of this.

transgressive resistance: Type of resistance attained from crosses between cultivars classified as susceptible to pathogens. Transgressive segregation of increased level is likely to be durable if it was derived from parents that have shown durable resistance.

transgressive segregation: The phenomenon through which we get variation in F_2 or later generations outside the range of both the parents.

transgressive variation: The production of F_2 phenotypes that exceed the parental extremes.

***trans*-heterozygote:** A heterozygote of two linked genes with a mutant and wild-type allele linked, a+ / +b. This is a more appropriate term than "repulsion phase" of linkage, since like alleles do not repel.

***trans*-hydrogenation:** Sequential transfer of electrons in cellular oxidation cycles.

transient diploid: The stage of the life cycle of predominantly haploid fungi (and algae) during which meiosis occurs.

transient expression: A technique in which DNA is introduced into eukaryotic cells, and its transcription is analyzed after it has reached the nucleus, but before it has integrated into the genome.

transient peptide: The short, leader sequence cleaved from proteins that are imported into cellular organelles by posttranslational passage of the membrane.

transient polymorphism: Some populations contain detrimental mutations that are gradually eliminated by selection. Selection in such cases acts to decrease polymorphism.

transition mutation: A point mutation that results in a purine being replaced by another purine or a pyrimidine by another pyrimidine. (Freese, 1959)

translation: The synthesis of a polypeptide, the amino acid sequence of which is determined by the nucleotide sequence of an mRNA in accordance with the rules of the genetic code. Also referred to as *cytoplasmic protein synthesis.*

translational amplification: The use of the same mRNA molecule many times, allowing the synthesis of large amounts of polypeptide from a single mRNA and therefore, from a single gene product.

translational control: Control mechanisms that operate at the level of translation.

translocase (EF-G): Elongation factor in prokaryotes necessary for proper translocation at the ribosome during the translation process. Replaced by eEF2 in eukaryotes.

translocation: The movement of a ribosome from one codon to the next on an mRNA molecule during translation.

translocation of chromosome: A rearrangement in which part of a chromosome is detached by breakage and then becomes attached to some other chromosome.

translocation of gene: The appearance of a new copy at a location in the genome elsewhere from the original copy.

translocation of protein: The movement of a ribosome from one codon to the next on an mRNA molecule during translation.

translocation of ribosome: The movement of a ribosome one codon along mRNA after the addition of each amino acid to the polypeptide chain.

translocation shift: The transfer of an interior segment of a chromosome to the interior of a nonhomologous chromosome.

translocation tester set: A set of a minimum number of translocation homozygous stocks. Each involves only two pairs of chromosomes in such a way that each chromosome is represented at least once in any one of these stocks. One study tester set should be able to identify the chromosomes involved in any unknown interchange of the same species.

transmission genetics: The study of the mechanisms involved in the passage of a gene from one generation to the next.

transplacement of gene: A technique, originally developed in yeast molecular biology, for exchanging a DNA sequence on a recombinant plasmid for the equivalent sequence in one of the host's chromosomes. Thus, a wild-type copy involves specific protein "carriers" located within the cell membrane.

transposable genetic element: In prokaryotes and eukaryotes, any of a class of diverse DNA segments that can insert into nonhomologous DNA, exit, and relocate in a reaction that is independent of the general recombination function of the host. Prokaryotic transposable elements are of three types: (a) insertion sequences (IS elements) that are of size 0.8 to 1.5 kb that encode only transposition determinants; (b) transposons (TN) that are greater than 5 kb long and encode additional transposition determinants (such as antibiotic resistance) and are often bracketed by IS elements; and (c) transposable phages that carry out site-specific recombination reactions which neither require DNA replication nor degradation or resynthesis of DNA. Eukaryotic genetic elements fall into three major classes: (a) classical transposons; and (b) retrovirus or retroviral-like retroposons; and (c) retroposons. (McClintock, 1951; Cohen, 1976)

transposase: The enzyme that catalyzes transposition of a transposable genetic element. (Berg, 1977)

transposition: The process whereby a transposon or insertion sequence inserts into a new site on the same or another DNA molecule. The exact mechanism is not fully understood, and different transposons may transpose by different mechanisms. Transposition in bacteria does not require extensive DNA homology between the transposon and the target DNA. The phenomenon is therefore described as *illegitimate recombination.*

transposition immunity: The ability of certain transposons to prevent others of the same type from transposing to the same DNA molecule.

transposition pseudoallele: Pseudoalleles are very closely linked. Same as TRANSHETEROZYGOTE.

transposon: A segment of DNA, which generally consists of more than 2,000 nucleotides, that is capable of moving into and out of a chromosome or plasmid and/or from place to place within the chromosome in both prokaryotes and eukaryotes. It is capable of turning genes "on" and "off." *See* TRANSPOSABLE GENETIC ELEMENT. (Hedges and Jacob, 1974)

transposon tagging: The blocking activity of a functional gene by inserting a foreign DNA sequence. (Bingham, Levis, and Rubin, 1981)

transposon-mediated gene transfer: Transferring a gene from one species to another by use of a transposon.

***trans*-splicing:** Intermolecular splicing of exons on different transcripts, as opposed to *cis*-splicing which involves intramolecular splicing.

transtranslation: A phenomenon where an RNA molecule (10Sa RNA) acts both as transfer RNA, which is responsible for transferring amino acids to the growing peptide, and as messenger RNA, in *E. coli*. (Atkins, 1996)

transvection: The ability of a locus to influence the activity of an allele on the other homolog only when two chromosomes are synapsed. (Lewis, 1954)

transversion: A point mutation that results in a purine being replaced by a pyrimidine, or vice versa. (Freese, 1959)

triangle code: The discredited idea that RNA organizes itself into triangular shapes into which amino acids fit.

trihybrid: An individual heterozygous for three pairs of alleles.

triisosomic: An individual that is lacking one chromosome pair but has three additional homologous isochromosomes for the same arm of the missing pair. (Kimber and Sears, 1968)

triplet code: A group of three successive nucleotide in RNA or DNA that, in the genetic code, specifies a particular amino acid in the synthesis of polypeptide chains.

triplet binding assay: An experimental technique that enables the coding specificity of a triplet of nucleotide to be determined.

triplex: The condition in an autotetraploidy in which three genes are dominant and one recessive. (Blakeslee, Belling, and Farnham, 1923)

triplex DNA: *See* DNA HELIX H.

triplicate genes: Three genes, any one of which produces the same phenotypic effect, e.g., red, seed-coat color in wheat may be produced by R_1, R_2, or R_3.

triploid: A polyploid cell, tissue, or organism with three sets of chromosomes (3n). (Němec, 1910)

trisomic: An individual with one or more extra chromosomes in a diploid set (2n + 1; 2n + 1 + 1). (Blakeslee, 1921)

trisomy: The diploid condition plus one extra chromosome.

tritelosomic: An individual that is lacking one chromosome pair but has three additional, homologous, telocentric chromosomes. (Kimber and Sears, 1968)

triticale: An allopolyploid obtained by combining the chromosomes of wheat and rye to produce a new species, *Triticosecale wittmack.*

tritium: A radioactive isotope of hydrogen.

trivalent: Three, synapsed, homologous chromosomes.

tRNA: *See* TRANSFER RNA.

tRNA deacylase: The enzyme responsible for cleaving the bond between a tRNA molecule and the growing polypeptide during translation.

tRNA isoacceptors: Differing species of tRNA that bind the same amino acid but, because of coding degeneracy, have differing anticodon triplets. *See* DEGENERATE CODE.

tRNA nucleotidal transferase: The enzyme responsible for the post-transcriptional attachment of the nucleotide sequence 5′-CCA-3′ to the 3′ end of a tRNA molecule.

tRNA precursor: A form of tRNA that is about 20 percent larger than the mature molecules and contains extra sequences at both the 5′ and the 3′ ends. In prokaryotes, these are removed by specific enzymes, RNase P and RNase Q, which cleave at the 5′ and 3′ ends, respectively.

tRNA releasing factor: A factor catalyzing the release from ribosomes. It is distinct from the G factor in that it loses its activity upon heating and does not require GTP for its action. It may play a role in removing tRNA from the ribosome after the polypeptide-chain termination factor has released the completed chain from the ribosome-bound tRNA. (Ishitsuka and Kaji, 1970)

true breeding: The organisms are homozygous for the trait under consideration.

true heritability: *See* HERITABILITY.

true overdominance: Heterosis ($F_1 > P$) for a single locus. In contrast, pseudo-overdominance may occur due to repulsion-phase linkage of more than one gene/loci mimicking the true overdominance, while superdominance relates to heterosis accruing from heterozygotes that carry divergent alleles, e.g., $A_1A_4 > A_1A_3 > Q_1A_2$.

truncate distribution: An atypical sample of a population in which certain types of matings or individuals have been omitted.

truncation selection: A breeding technique in which individuals in whom quantitative expression of a phenotype is above or below a certain value (the truncation point) are selected as parents for the next generation.

tumor: Aggregates of cells derived from an initial aberrant cell that although surrounded by the normal tissue, is no longer integrated into the environment.

tumor suppressor genes: Those genes which code for proteins that inhibit cell replication.

tumorigenesis: A multiple process in which a normal cell progresses in a step-by-step fashion to a fully malignant tumor cell.

tumor-inducing (Ti) plasmid: A 140 to 235 kb plasmid found in *Agrobacterium tumefaciens* having a virulence trait.

twin species: Two species that are phenotypically very similar but separated from each other by reproductive isolation. *See* SIBLING SPECIES.

twin spot: A pair of mutant sectors within wild-type tissue, produced by a mitotic crossover in an individual of appropriate heterozygous genotype.

twisting number: In double helical DNA, this is the number of base pairs divided by the number of base pairs per turn of the double helix.

two-dimensional electrophoresis: A method for enhancing the separation of proteins and other macromolecules that are of similar size and charge and not resolved by electrophoresis in one dimension.

two-dimensional paper chromatography: *See* PAPER CHROMATOGRAPHY.

two-out-of-three reading frame: The hypothesis according to which codons may be read by relying mainly on the Watson-Crick pairs formed with the first two codon positions; mispaired nucleotides in the third codon and anticodon, wobble positions make only a marginal contribution to the total stability of the reading interactions. (Lagerkvist, 1978)

two-point test cross: A test cross involving two loci.

two-step ligation: A gene-cloning protocol in which linearized vector molecules and insert DNA fragments are mixed and ligated at high DNA concentration. This favors the joining of vector and insert DNA. The DNA solution is then diluted for the second step of the ligation. This favors the circularization of recombinant molecules.

two-strand double crossover: A double crossover that involves only two of the four chromatids of a tetrad.

Ty element: In yeast, any of three related families of dispersed, transposable, genetic elements (Ty=transposon yeast) that share considerable structural and functional features with retroviruses. They contain terminal direct-repeated sequences of ~335 bp that share extensive homology at the nucleotide sequence level and transpose by a reverse transcriptase-mediated process. Ty elements transpose preferentially into the regulatory regions at the 5′ ends of yeast genes. (Cameron, Loh, and Davis, 1979)

type I callus: A type of adventive embryogenesis found with gramineous monocots, which has been induced on an explant where the somatic embryos are arrested at the coleptilar or scutellar stage of embryogeny. The embryos are often fused together, especially at the coleorhizal end of the embryo axis. The tissue can be subcultured and maintains this morphology.

type II callus: A type of adventive embryogenesis found with gramineous monocots, which has been induced on an explant where the somatic embryos are arrested at the globular stage of embryogeny. The globular embryos often arise individually from a common base. The tissue can be subcultured and maintains this morphology.

type I error: In statistics, the rejection of a true hypothesis.

type II error: In statistics, the accepting of a false hypothesis.

type number: The most frequently occurring chromosome number in a particular taxonomic group (= modal number). (Harvey, 1917)

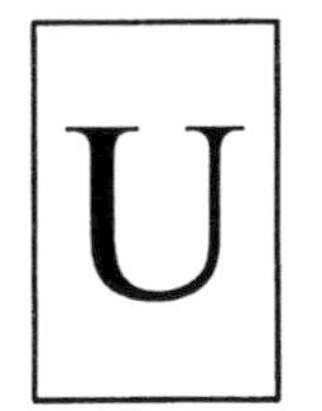

U orientation: Descriptive of the situation in which the vector and the inserted fragment of DNA have the opposite orientations.

U protein: Hypothetical protein thought to regulate the transition of cells from G_0 to G_1 phase of the cell cycle and thus inevitably into S phase. The idea would be that the concentration of this unstable (U) protein would have to exceed a threshold level for triggering progression through the cycle, and that this would only happen if the cell had adequate access to growth factors or to nutrients. Also known as *trigger protein*.

U snRNPs: Small nuclear particles composed of U snRNAs and protein. These are involved in the cleavage of primary transcript at the 3′ end, splicing of introns and addition of poly(A) tails at the 3′ end of RNA transcripts.

U1 to U11 snRNAs: A family of stable, uridine-rich, small, nuclear RNA molecules originally found in rat Novikoff hepatoma cells but subsequently discovered in a range of mammals and other higher eukaryotes. These RNAs are complexed with 7 to 8 proteins to form small nuclear ribonucleoprotein particles (snurps). They contain several modified bases, have no 3′ poly(A) tail, and mostly contain a 5′ trimethylguanosine cap structure. In mammals, at least 11 U snRNA species are known. U1 to U7 and perhaps U11 are directly involved in maturation and processing of cellular RNAs. U1 to U6 are classified as major U-type snRNAs, which have 10^5 to 10^6 copies. U7 to U11 are classified as minor U-type snRNAs, which have 2.5×10^4 copies. All U-type RNAs function through specific base-paired hydrogen bonding with other RNAs.

UACUAAC box: In yeast introns, a conserved sequence that specifies the branch point for the lariat and to which U2 snRNA binds by complementary base pairing during RNA splicing.

ubiquitin: Acidic protein widely, and perhaps universally, distributed among eukaryotes. It binds strongly to chromatin and may have a structural role in chromatin conformation. Binds specifically to histone H2A by a covalent linkage; the complex has come to be known as protein A24. (Schlesinger, Goldstein, and Niall, 1975)

ultracentrifuge: A machine used to subject samples to a high gravitational force by spinning them at a very high speed, up to 60,000 rpm, and a centrifugal field up to 500,000 times gravity. It is capable of rapidly sedimenting macromolecules.

ultraviolet (UV): Nonionizing radiations produced from mercury vapor lamps or tubes and used for induction of mutation in lower organisms. UV has a spectrum beyond the violet end of visible light (wave length < 400 nm).

ultraviolet fluorescent microscopy: Ultraviolet microscope in which the ultraviolet light is directed on biological material that fluoresces in a particular color when so irradiated. This may be because of natural fluorescence or because of staining in or association with fluorescently tagged antibody. *See* FLUORESCENCE MICROSCOPY.

ultraviolet spectroscopy: The determination of the amount of ultraviolet radiation absorbed by a sample, used to identify chemical compounds and to measure the concentrations of nucleic acid solutions.

ultraviolet visualization: A technique used to detect the position of proteins, nucleic acids, and other compounds that absorb ultraviolet light, or fluoresce in the presence of ultraviolet light, on paper, gel or thin layer chromatograms, or on electrophoretic gels. In some cases, the matrix is sprayed with a suitable reagent prior to inspection to enable visualization of the compounds.

μm (micrometer): 1×10^{-6} meter.

unassigned reading frame: A gene-like nucleotide sequence with proper initiator and terminator codons but with no known function. Intronic URFs in fungal mitochondiral genes probably encode genes required for splicing of these introns from precursors of mRNAs.

unbalanced translocation: Chromosome translocation in which some loss of chromosomal material has resulted or in which a deleterious genetic effect is evident following the translocation.

uncoiling of chromomeres: Situation most readily seen in the puffing of polytene chromosomes and the looping-out of chromatin loops in lampbrush chromosomes.

uncondensed chromatin: Chromatin that is not highly compacted and is therefore more likely to be transcriptionally active; there is a strong correlation between gene activity and chromatin decondensation. Both the puffs and interbands of polytene chromosomes are uncondensed and are sites of intense and moderate transcriptional activity, respectively.

underdominance: A phenotypic relation in which the phenotypic expression of the heterozygote is less than that of either homozygote.

underwinding: Underwinding of DNA is produced by negative supercoiling (because the double helix is itself coiled in the opposite sense from the intertwining of the strands).

undifferentiated: With plant cells, this exists in a state of cell development characterized by isodiametric cell shape, very little or no vacuole, a large nucleus, and is exemplified by cells comprising an apical meristem or embryo. With animal cells, this is a state wherein the cell in culture lacks the specialized structure and/or function of the cell type *in vivo*.

unequal bivalent: *See* BIVALENT, UNEQUAL.

unequal crossing-over: Nonreciprocal crossing-over during meiosis caused by mismatching of homologous chromosomes. This usually occurs in regions of tandem repeats.

unidentified reading frame: An open reading frame recognized from a DNA sequence for which no genetic function is known.

unidirectional replication: The movement of a single replication fork from a given origin.

unilinear inheritance: The transfer of a donor chromosome to a suitable recipient (bacteria).

unineme hypothesis: The concept that an unduplicated chromosome, one prior to the S phase of DNA replication, has a single molecule of DNA duplex running from end to end. The concept is now generally accepted, although it is clear that the molecule is highly twisted upon itself both by reason of packing around nucleosomes and packing of nucleosomes in the 30 nm fibers of chromatin, and further packing of these fibers relative to the chromosomes' axis. *Compare* POLYNEME HYPOTHESIS.

uninemic chromosome: A chromosome consisting of one double helix of DNA.

uniparental inheritance: An inheritance pattern where the offspring have received certain phenotypes from only one parent. This inheritance pattern is due to transmission of DNA containing cytoplasmic particles.

unipartite structures: Single units.

unique DNA: A length of DNA with no repetitive nucleotide sequences. Also known as *single copy DNA*.

unisite mutant allele: A mutant allele differing from its wild-type form at only one site.

unit character: A character or trait controlled by one pair of alleles, as opposed to quantitative or polygenic characters. (Castle, 1966)

unit factor: The term used by Mendel for the gene.

univalent: A single chromosome seen segregating in meiosis.

universal codon: A triplet codon codes for the same amino acid in diverse forms of life, e.g., bacteria, wheat, *Drosophila,* humans.

universal nucleoside: 1-(2′-deoxy-ß-D-ribofuranosyl)-3-nitropyrrole, designated "M" is a universal nucleoside that maximizes stacking while minimizing hydrogen-binding interactions without sterically disrupting a DNA helix. The universal nucleoside is used at ambiguous sites in DNA primers. (Nichols, 1994)

unmatched S gene: Any S gene that is represented either in the style or pollen but not in both, in a given pollination. (An S gene is any gene controlling the specificity of incompatibility reactions in plants.)

unmixed families: Groups of four codons sharing their first two bases and coding for the same amino acid.

unreduced gamete: A gamete formed by abnormal division where meiosis II had not taken place.

unscheduled DNA synthesis: Synthesis that occurs at times other than the S phase of the cell cycle. Most of such synthesis is DNA repair, although synthesis of mitochondrial DNA in the cytoplasm is also continuous. (Rasmussen and Painter, 1964)

unstable equilibrium: A frequency equilibrium between alleles in a population that is readily upset because of a selective disadvantage of the heterozygotes.

unstable gene: A gene that mutates frequently.

unstable mutation: A mutation that has a light frequency of reversion; a mutation caused by the insertion of a controlling element, whose subsequent exit produces a reversion.

untranscribed spacer: DNA sequence that lies between two-coding sequences and is not transcribed by RNA polymerase. Although such sequences do exist, some identified in the past (e.g., the spacers between the multiple copy of rRNA genes) are known to be transcribed.

untranslated sequence: A region of mRNA that is not used in the synthesis of an amino acid sequence of a given peptide or protein. They usually appear at either end of the sequence that codes for the amino acid sequence.

untwisting enzyme: An enzyme that preserves or restores the helical structure of double-stranded DNA during replication of circular DNA molecules. These enzymes produce a nick, allowing free rotation, and then reseal the nicked DNA strand once the supercoils have been removed.

unusual bases: Other bases, in addition to adenine, cytosine, guanine, and uracil, found primarily in tRNAs.

unusual chromosomes: The chromosomes that show adaptational forms of the normal chromosomes (e.g., lampbrush chromosomes, polytene chromosomes), or these may be permanently specialized structures (e.g., B chromosomes, sex chromosomes).

unwinding enzyme: A protein that binds to DNA strands at the growing point of the replication fork during replication of double-stranded DNA.

unwinding protein: Protein that untwists the DNA double helix before the replication fork reaches that region during DNA synthesis. Such a protein, discovered in phage T_4, has been termed the gene 32 protein.

up promoter mutation: Mutation in the promoter that increases the frequency of initiation of transcription.

upstream: Toward the 5′ end of a polynucleotide.

upstream activating sequence (UAS): A DNA segment required for activating transcription of yeast genes (Guarente, 1984; Siliciano and Tätchell, 1986)

uracil: 2, 6-dioxypyrimidine, one of the nitrogenous bases found in RNA.

uridine: The ribonucleoside that contains the pyrimidine uracil.

uridylic acid: The ribonucleotide that contains the pyrimidine uracil.

U-RNA: A nuclear RNA molecule involved in splicing and other transcript-processing events.

U-RNP: A nuclear particle, consisting of one or two U-RNAs and several proteins, involved in splicing and other transcript-processing events.

use and disuse doctrine: The discredited idea that the more a structure is used, the more prominent it becomes in future generations, and that the less it is used, the less prominence it assumes in later generations. Also known as *doctrine of acquired characteristics*.

UV endonuclease: An endonuclease capable of cleaving a DNA chain at the 5′ side of a thymine dimer.

variable: A property that may have different values in various cases.

variable number of tandem repeat (VNTR) loci: Loci that are hypervariable because of tandem repeats. Presumably, variability is generated by unequal crossing-over.

variance, environmental: The variance resulting from environmental or nongenetic causes.

variance, genetic: The variance resulting from genetic causes.

variance, phenotypic: The total variance—the sum of the environmental and the genetic variance.

variant: (1) An individual organism that is recognizably different from an arbitrary standard type in that species; (2) a culture exhibiting a stable phenotypic change whether genetic or epigenetic in origin.

variate: A specific numerical value of a variable.

variation: The difference among individuals of a species in a population.

variegation: Patchiness; a type of position effect that results when particular loci are continuous with heterochromatin. (Schultz, 1936)

varietal revamping: Reestablishment of a variety (released one) that has run out of cultivation (deteriorated due to various reasons). It can be rectified by secondary selection or other means and then again can be revamped under cultivation.

variety: A subdivision of a species. An agricultural variety is a group of similar plants that by structural features and performance can be identified from other varieties within the same species.

variety blend: A mechanical mixture of seed of two or more varieties.

V-chromosome: Chromosomes with two arms, i.e., mediocentric chromosomes.

vector: A DNA molecule capable of replication into which a gene is inserted by recombinant DNA techniques. Artificial vectors are constructed by cutting and pasting DNA with enzymes. Some vectors (expression vectors) are specifically designed to promote the transcription of a particular gene.

vegetative cell: A cell that is not a sex cell; i.e., one that divides by mitosis.

vegetative propagation: Reproduction of plants using a nonsexual process involving the culture of plant parts, such as stem and leaf cuttings.

vehicle: The host organism used for the replication or expression of a cloned gene or other sequence. *Compare* VECTOR, the DNA molecule that contains the cloned gene. The term is little used and is often confused with vector.

vehicle plasmid: A plasmid containing a piece of passenger DNA; it is used in recombinant DNA work.

velocity sedimentation analysis: A method that measures the mass of a molecule or particle by determining the rate at which it sediments through a dense solution when subjected to a high centrifugal force.

vernalization: The treatment of seeds before sowing to hasten flowering. Vernalization may be accomplished in certain species by exposure of germinating seeds to temperatures slightly above freezing.

vertical resistance (VR): Resistance against a single race of pathogen. It is known to be controlled by major genes that have large effects and reveal gene-for-gene relationship. Also known as *race-specific resistance*. *See* GENE-FOR-GENE HYPOTHESIS. (Vanderplank, 1963)

vertical transmission: The transmission of traits controlled by major genes that have large effects and reveal gene-for-gene relationship. For example, race-specific resistance shows vertical transmission.

Vertifolia effect: The masking of horizontal resistance (HR) alleles while selecting for vertical resistance (VR) as happens in the variety *Vertifolia* of potato.

very short patch (VSP) repair: In *E. coli,* a repair mechanism that repairs T/G mismatches regenerating original CC(AT)GG sites. The mechanism has probably evolved to minimize C→T transitions that result from deamination of methylated cytosines. (Leib, 1966)

vilmorin principle: Selection carried out on the basis of progeny test (plant-to-row performance). Pure-line breeding is based on this principle.

viral chromosome: *See* CHROMOSOME.

virion: A complete virus particle consisting of its nucleic acid core and protein coat. (Lwoff, Anderson, and Jacob, 1959)

virogene: The complete viral genome.

viroid: Short strands of RNA capable of causing disease in plants. (Diener, 1971)

virulence: The ability to produce disease in any particular instance.

virulent phage: A bacterial virus that typically destroys the host cell it infects.

virus-free: Free from specified viruses based on tests designed to detect the presence of the organisms in question.

visibles: Mutations recognizable by their phenotypic expression, in contrast to vitality mutations and lethal factors.

vitality mutation: Any mutation the effect of which, in contrast to visibles and lethal factors, is difficult to determine but that changes the viability of the carrier genotype if present in an effective dosage.

vitrification: Abnormal looking regenerates of shoot cultures.

vybrid: A cross between two open-pollinated varieties. Hybrid is a cross between two self-pollinated (inbred) lines, and cybrid is a somatic (parasexual) hybrid where cytoplasms of the two parents have been hybridized but the nuclear genome of one is eliminated.

W: The symbol representing adaptive value (or relative fitness) in mathematical treatments of natural selection.

Wahlund effect: A subdivided population contains fewer heterozygotes than predicted despite the fact that all subdivisions are in Hardy-Weinberg proportions. (Wahlund, 1928)

Wallace effect: Selection for reproductive isolation in the areas of sympatry based on a Hardy-Weinberg equilibrium, when two divergent populations—previously isolated and subjected to forces differentiating allele frequencies between them—occur sympatrically and are sampled as a single population. (Grant, 1966)

Watson-Crick base pairing: The normal base pairing, e.g., A with T and G with C in DNA and A with U and G with C in RNA.

Watson-Crick model: Double-helical, right-handed model for DNA structure which proposed that two polynucleotide chains were associated by specific hydrogen bonds between complementary base pairs with the sugar phosphate backbone forming a double helix and thus, providing a rigid framework of bases. Although alternative structures for DNA (e.g., left-handed Z-form) have been discovered subsequently, the model has proved correct and has provided enormous stimulus and insight in the field of molecular genetics. (Watson and Crick, 1953)

weight: A differential value assigned to an estimate of a quantity, relative to other estimates of the same quantity, for the purpose of combining the estimates.

Western blotting: A technique for probing for a particular protein using antibodies. *See* SOUTHERN BLOTTING.

wide hybridization: Interspecific and intergeneric hybridization that is a first step to introducing alien variation and to transferring desirable genes and traits from wild species into cultivated species. Barriers to wide hybridization are incompatability between the parental species, viability of the F_1 hybrids, and sterility of the F_1 hybrids or their progeny.

wild-type: A gene, cell, or organism that displays the typical phenotype and/or genotype for the species and is therefore adopted as a standard.

wobble base pairing: The pairing of an mRNA codon with a tRNA anticodon in which the first two bases of a codon have normal pairing and the third base has abnormal base pairing. (Crick, 1966) The third position deviations consist of pairing between the bases shown in Table 2.

TABLE 2. Wobble Base Pairing

5′ end of anticodon (third position)	3′ end of codon
G	U or C
C	G only
A	U only
U	A or G
I	U, C, or A

world collection: Synonymous with germplasm collection. *See* GERMPLASM COLLECTION.

Wright's coefficient: *See* INBREEDING COEFFICIENT.

Wright's effect: Nonadaptive differentiation into different types due to "drift" followed by random fixation in small isolated populations.

Wright's inbreeding coefficient: The proportion of loci at which an individual organism is homozygous.

writhing number: The number of times a duplex axis crosses over itself in space.

x: The basic number of chromosomes in a polyploid series.

X, 2X, 3X, Etc: Symbols for the number of genomes in the gametic nucleus ob species in a genus with a polyploid series.

X_1, X_2, X_3, Etc: Different generations obtained from an individual that has been irradiated.

xenia effect: The direct effect of male gamete on tissues other than embryonic ones. Xenia effect is useful in the study of gene dosage effect. (Focke, 1881)

Xma I: A type II restriction enzyme from the bacterium *Xanthomonas malvacearum* that recognizes the DNA sequences shown below and cuts at the sites indicated by the arrows:

```
       ↓
5′  C  C  C  G  G  G  3′
3′  G  G  G  C  C  C  5′
                   ↑
```

x-ray crystallography: A technique for deducing molecular structure by aiming a beam of x-rays at a crystal of the test compound and measuring the scatter of rays.

x-ray diffraction analysis: A method for determining the structure of a molecule by analyzing the angles at which x-rays are deflected by a crystal of the compound.

x-ray diffraction pattern: The pattern of spots on an x-ray sensitive film that is produced by passing x-rays through a crystal.

Y

Yates' correction: A correction applied in the calculation of normal deviates or χ^2 values to allow for the discrepancy arising by the observations being discontinuous, while tables of the normal deviate and χ^2 are calculated on the supposition of continuity in the variate.

yeast artificial chromosome (YAC): A cloning vector comprising the structural components of a yeast chromosome and able to clone very large pieces of DNA.

yeast centromere plasmid (YCp): A plasmid that carries a functional yeast centromere.

yeast episomal plasmid (YEp): Many strains of yeast contain a 2 μm (6 kb) long plasmid that has no known function.

yeast genome: *Saccharomyces cevervisiae* is a haploid eukaryote having 12,068 kb DNA consisting of 6,340 genes, including 5,885 potential protein-encoding genes, 140 rRNA genes, 40 snRNA genes, and 275 tRNA genes. (Goffeau et al., 1996)

yeast integrative plasmid (YIp): A yeast vector that relies on integration into a host chromosome for replication. It contains a yeast selectable marker inserted into bacterial plasmid vectors (e.g., pBR322).

yeast replicative plasmids (YRp): These plasmids contain autonomous replicating sequences (ARS) derived from yeast chromosomes.

yeast vectors: Yeast cells used as carriers of plasmids, in which novel genes may be replicated, or as transformed cells used to produce the products of integrated novel genes. Yeast vectors are capable of selection and maintenance in either yeast or *E. coli*.

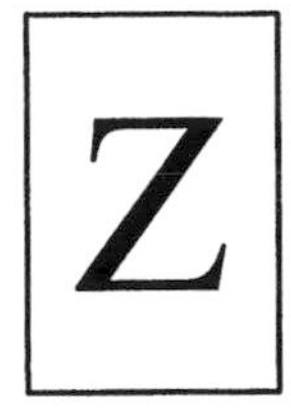

z: The natural logarithm of the ratio of two estimated standard deviations.

Z DNA: The DNA in which sugar and phosphate linkage follows a zig-zag pattern. Such DNA has a left-handed, double helical model. *See* DNA HELIX Z. (Wang et al., 1979)

zero selection: When selection intensity is the highest and the proportion of the population rejected is 100 percent (vice versa in the case of no selection), i.e., the lowest selection and no rejection at all. In this case the entire population is carried forward.

zero time-binding DNA: DNA that the duplex form at the start of a reassociation reaction; it results from intramolecular reassociation of inverted repeats.

zinc finger motif: A tandemly repeated, protein-sequence motif found in several DNA-binding proteins containing 28 to 30 amino acids each. Two cysteines and two histidines are present at invariant positions. These motifs are encoded by various regulatory genes (helix-turn-helix motif). Proteins containing zinc finger motifs participate in regulation of genetic transcription.

zone electrophoresis: A type of electrophoresis in which the mobile passes through a support medium such as paper, cellulose acetate, starch gel, silica gel, polyurethane foam, or acrylamide polymers. The sample is applied to the support in a narrow band approximately midway between the buffer reservoirs housing the electrodes. During electrophoresis, each component migrates toward the cathode or anode, depending on its charge, at a characteristic rate. Using this method, charged species that are separated into discrete units can be divided into sections and individual components may be separately eluted.

zoo blot: This describes the use of Southern blotting to test the ability of a DNA probe from one species to hybridize with the DNA from the genomes of a variety of other species.

Z RNA: A left-handed RNA double helix; the conformational transition from A-form to Z-form requires more extreme conditions than the transition of B to Z DNA. It has relevance in those cases in which double-stranded RNA occurs (e.g., ribosomes and some viruses). *See* DNA HELIX Z.

zygosome: A newly fused chromosome pair produced by synapsis (zygotene). Also known as *mixochromosome*.

zygote: The cell that results from the fusion of gametes during meiosis. (Bateson, 1902)

zygotene: A substage of prophase I of meiosis I during which homologous chromosomes begin to synapse. The points at which the chromosomes first come in contact with each other in pairing during zygotene are known as *contact points*.

zygotene DNA (ZYG DNA): A fraction of eukaryotic nuclear DNA that fails to replicate during premeiotic S-phase and undergoes replication during zygotene in coordination with chromosome pairing. (Hotta, Tabata, and Stern, 1984)

zygotene RNA: RNA transcribed from zygotene DNA. (Hotta et al., 1985)

zygotic checkerboard: *See* PUNNETT SQUARE.

zygotic induction: When a prophage is passed into an F^- cell during conjugation, it may begin vegetative growth.

zygotic lethal: A lethal gene whose effect is in the embryo, larva, or adult, in contrast to the gametic lethal affecting a gamete.

zygotic mutation: A mutation occurring in the zygote shortly after fertilization.

zygotic selection: The forces acting to cause differential mortality of an organism at any stage (other than gametes) in its life cycle.

zymogram: Electrophoretic gels used to reveal and compare protein polymorphism, as of isoenzymes such as lactate dehydrogenase subunits.

Bibliography

Abercrombie, M. and Haeysman, J.E.M. (1953). "Observations on the social behavior of cells in tissue culture I. Speed of movement of chick heart fibroblasts in relation to their contacts," *Experimental Cell Research,* 5:111-131.

Adhya, S., Gottesman, M., and de Crombrugghe, B. (1974). "Release of polarity in *Escherichia coli* by gene N of phage λ: Termination and antitermination of transcription," *Proc Natl Acad Sci USA,* 71:2534-2538.

Adolph, K.W., Cheng, S.M., Paulson, J.R., and Laemmi, U.K. (1977). "Isolation of a protein scaffold from mitotic He La cell chromosomes," *Proc Natl Acad Sci USA,* 74:4937-4941.

Agar, W.E. (1911). "The spermatogenesis of *Leptosiren paradoxa,*" *Quartl J Micr Sci,* 67:1-13.

Air, G.M., Blackburn, E.H., Coulson, A.R., and Galibert, F. (1976). "Gene F of bacteriophage ϕX174. Correlation of nucleotide sequences from the DNA and amino acid sequences from the gene product," *J Mol Biol,* 107:445-458.

Alberts, B.M., Bray, D., Lewis, J., Raff, M., Roberts, K., and Watson, J.D. (1989). *Molecular Biology of the Cell* (second edition). New York: Garland, pp. 692-713.

Alberts, B.M. and Frey, L. (1970). "T4 bacteriophage gene 32: A structural protein in the replication and recombination of DNA," *Nature,* 227:1313-1318.

Alexeiff, A. (1917). "Sur la fonction glycoplastique due kinetoplaste (Kinetonucleus) chez las flagelles," *Soc Biol Paris,* 80:512-530.

Altmann, R. (1889). "Über Nuckleinsäuren," *Arch Anat Physiol Lpz Physiol Abt,* 524-527.

Ames, B.N. (1971). The detection of chemical mutagens with enteric bacteria. In *Chemical Mutagens*, Vol. I, ed. Hollaender, A., New York: Plenum Press, p. 267.

Ames, B.N. and Garry, B. (1959). "Coordinate repression of the synthesis of four histidine biosynthetic enzymes by histidine," *Proc Natl Acad Sci USA,* 45:1453-1461.

Ames, B.N. and Hartman, P.E. (1963). "The histidine operon," *Cold Spr Harb Sym,* 28:349-356.

Anderson, E. (1953). "Introgressive hybridization," *Biol Rev,* 28:280-307.

Anderson, E. and Hubricht, L. (1938). "Hybridization in *Tradescantia*. The evidence for introgressive hybridization," *Am J Bot,* 25:396-402.

Angel, T., Austin, B., and Catcheside, D.G. (1970). "Regulation of recombination at the *his-3* locus in *Neurospora crassa,*" *Aust J Biol Sci,* 23:1229-1240.

Astbury, W.T., Beighton, E., and Weibull, C. (1955). "The structure of bacterial flagella," *Symp Soc Exp Biol,* 9:282-305.

Athwal, R.S. and McBridge, O.W. (1977). "Serial transfer of a human gene to rodent cells by sequential chromosome-mediated gene transfer," *Proc Natl Acad Sci USA,* 74: 2943-2947.

Atkins, J.E. (1996). "Genetic code: A case for *trans* translation," *Nature,* 379:769.

Avery, D.T., MacLeod, C.M., and McCarthy, M. (1944). "Studies on the chemical nature of the substance inducing transformation of pneumococcal types. Induction of transformation by a deoxyribonucleic acid fraction isolated from *Pneumococcus* Type III," *J Exp Med,* 79:137-158.

Baltimore, D. (1970). "Viral RNA-dependent DNA polymerase," *Nature,* 226: 1209-1211.

Bangham, A.D., Standish, M.M., and Watkins, J.C. (1965). "Diffusion of univalent ion across the lamellae of swollen phospholipids," *J Mol Biol,* 13:238-252.

Barrell, B.G., Air, G.M., and Hutchison, C.A., III. (1976). "Overlapping genes in bacteriophage φX174," *Nature*, 264:33-41.

Barski, G., Sorieul, S., and Cornefert, F. (1960). "Production dans descultures in vitro de deux souches cellulaires en association, de callules de caractere 'hybride,'" *Acad Sci Paris*, 251:1825-1827.

Bateson, W. (1894). *Materials for the Study of Variation*. London: MacMillan.

Bateson, W. (1902). *Mendel's Principles of Heredity.* Cambridge, MA: Cambridge Univiversity Press.

Bateson, W. (1905). In a letter to Sedgewick, A. from Bateson, W. (1928). "*Essays and Addresses*," ed. Bateson, B., Cambridge, MA: Cambridge University Press.

Bateson, W. (1907). "The progress of genetics since the rediscovery of Mendel's paper," *Progr Rei Botanicae,* l:368-382.

Bateson, W. and Punnett, R.C. (1906). "Experimental studies in the physiology of heredity," *Reports to the Evolution Committee of the Royal Society, III.* London: Harrison and Sons, p. 238.

Bateson, W. and Saunders, E.R. (1902). "Experimental studies in the phylogeny of heredity," *Rep Evolut Comm Roy Soc Rep I.* London: Harrison and Sons.

Bateson, W., Saunders, E.R., and Punnett, R.C. (1905). "Experimental studies in the phylogeny of heredity," *Rep Evolut Comm Roy Soc Rep II.* London: Harrison and Sons.

Bateson, W., Saunders, E.R., and Punnett, R.C. (1908). "Experimental studies in the phylogeny of heredity," *Rep Evolut Comm Roy Soc Rep IV.* London: Harrison and Sons.

Battaglia, E. (1963). Apomixis. In *Recent Advances in the Embryology of Angiosperms,* ed. Maheshwari, p. 249.

Bautz, E.K., Bautz, F.A., and Dunn, J.J. (1969). "*E. coli* σ factor: A positive control element in phage T4 development," *Nature*, 223:1022-1024.

Beadle, G.W. (1932). "A possible influence of the spindle fiber on crossing over in *Drosophila*," *Proc Natl Acad Sci USA,* 18:160-165.

Beadle, G.W. and Tatum, E.L. (1941). "Genetic control of biochemical reactions in *Neurospora*," *Proc Natl Acad Sci USA,* 27:499-506.

Becker, M.M. and Wang, J.C. (1984). "Use of light for footprinting DNA *in vivo*," *Nature*, 309:682-687.
Bělăr K. (1928). *Die zytologischen Grundlagen der Vererbung Hdb Vererb I.* Berlin: Borntraeger.a
Bell, P.B. and Stillman, B. (1992). "ATP-dependent recognition of eukaryotic origins of DNA replication by a multiprotein complex," *Nature*, 357:128-134.
Belling, J. (1925). "Genetics of sex in *Funaria hygrometrica:* A correction," *J Genet*, 15:245-266.
Belling, J. (1927). "The attachment of chromosomes at the reduction division in flowering plants," *Jour Genet*, 18:177-205.
Belling, J. and Blakeslee, A.F. (1926). "On the attachment of nonhomologous chromosomes at the reduction division in certain 25-chromosome Daturas," *Proc Natl Acad Sci USA*, 12:7-11.
Benda, C. (1898). "Ümber die spermatogenese der Vertebraten and höhera Evertebraten. II. Tail. Die Histogenese der spermein, Arch Anat *Physiol*," *Physiol Abtlg Leipzig*, pp. 393-415.
Bender, W. Spierer, P., and Hogness, D.S. (1983). "Chromosome walking and jumping to isolate DNA from *Ace* and *rosy* loci and the biothrax complex in *Drosophila melanogaster*," *J Mol Biol*, 168:17-33.
Benzer, S. (1955). "Fine structure of a genetic region in a bacteriophage," *Proc Natl Acad Sci USA*, 41:344-354.
Benzer, S. (1957). The elementary units of heredity. In *The Chemical Basis of Heredity*, eds. McElroy, W.D. and Glass, B. Baltimore: Johns Hopkins Press.
Benzer, S. (1961). "On the topography of genetic fine structure," *Proc Natl Acad Sci USA*, 47:403-416.
Berg, D. (1977). *DNA Insertion Elements, Plasmids and Episomes*, ed. Bukhari, A.I. et al. New York: Cold Spring Harbor Laboratory, p. 205.
Berget, S.M., Moore, C., and Sharp, P.A. (1977). "Spliced segments at the 5′ terminus of adnovirus2 late mRNA," *Proc Natl Acad Sci USA*, 74: 3171-3175.
Bingham, P.M., Levis, R., and Rubin, G.M. (1981). "Cloning of DNA sequences from the *white* locus of *D. melanogaster* by a novel and general method," *Cell*, 25:693-704.
Birdsell, J.B. (1950). "Some implications of the genetical concept of race in terms of spatial analysis," *Cold Spr Harb Symp*, 15:259-314.
Birnsteil, M., Speirs, J., Purdom, I., Jones, K., and Loening, U.E. (1968). "Properties and composition of the isolated DNA satellite of *Xenopus laevis*," *Nature*, 219:454-463.
Blakeslee, A.F. (1904). "Sexual reproduction in the Mucorinae," *Proc Am Acad Arts Sci*, 40:205-319.
Blakeslee, A.F. (1921). "Types of mutations and their possible significance in evolution," *Am Nat*, 55:254-267.
Blakeslee, A.F. (1927). "Nubbin, a compound chromosomal type in Datura," *Ann NY Acad Sci*, 30:1-5.
Blakeslee, A. F., Belling, J., and Farnham, M.E. (1923). "Inheritance in tetraploid *daturas*," *Bot Gaz*, 76:329-373.

Blobel, G. and Dobberstein, B. (1975). "Transfer of proteins across membranes. I. Presence of proteolytically procassed and unprocessed nascent immunoglobulin light chains on membrane-bound ribosomes of murine myeloma," *J Cell Biol,* 67:835-851.

Blum, B., Sturm, N.R., Simpson, A.M., and Simpson, L. (1991). "Chimeric gRNA-mRNA molecules with oligo(U) tails covalently linked at sites of RNA editing suggest that U addition occurs by transesterification," *Cell,* 65:543-550.

Bocke, J.D., Garfinked, D.J., Styles, C.A., and Fink, J.R. (1985). "*Ty* elements transpose through an RNA intermediate," *Cell,* 40:491-500.

Bogorad, L. (1975). "Evolution of organelles and eukaryotic genomes," *Science,* 188:891-898.

Bohlar, C., Nielsen, P.E., and Orgel, L.E. (1995). "Template switching between PNA and RNA oligonucleotides," *Nature,* 376:578-581.

Bollum, F.J. (1960). "Ologodeoxyribonucleotide primers for calf thymus polymerase," *J Biol Chem,* 235:PC18-PC2O.

Botstein, D., White, R.L. Skolnick, M., and Davis, R.W. (1980). "Construction of a linkage map in man using restriction fragment length polymorphisms," *Amer J Hum Genet,* 69:201-205.

Bouton, A.H. and Smith, M.M. (1986). "Fine-structure analysis of the DNA sequence requirements for autonomous replication of *Saccharomyces cerevisial* plasmids," *Mol Cell Biol,* 6:2354-2363.

Boveri, T. (1895). "Über das Verhalten der centrosomen bei der Bebruchtung des Serigeleies mebst allgemeimen Bemerkungen über centrosomen und Vermandtes," *Verh Phys Med Ges Würzb,* 29:1-30.

Brenner, S., Barnett, L., Crick, F.H.C., and Orgel, A. (1961). "The theory of mutagenesis,"*J Mol Biol,* 3:121-124.

Brenner, S. and Beckwith, J.R. (1965). "Ochre mutants, a new class of suppressible nonsense mutants,"*J Mol Biol,* 13:629-637.

Brent, R. and Ptashne, M. (1985). "A eukaryotic transcriptional activator bearing the DNA specificity of prokaryotic repressor," *Cell,* 43:729-736.

Bridges, C.B. (1913). "Non-disjunction of sex chromosomes of *Drosophila,*" *J exp Zool,* 15:587-603.

Bridges, C.B. (1917). "Deficiency," *Genetics,* 2:445-465.

Bridges, C.B. (1919). "Specific modifies of eosin eye colour in *Drosophila melanogaster,*" *J exp Zool,* 28:337-384.

Bridges, C.B. (1923). "Aberrations in chromosome materials," *Scient Pap Second Int Congr Eugenics,* 1:76.

Bridges, C.B. (1932). "The suppressors of purple," *Z indukt Agstamm-u VererbLehre,* 60:207.

Bridges, C.B. (1934). "The testcross—A suggested genetic term," *J Hered,* 25:18.

Bridges, C.B. and Morgan, T.H. (1923). "The third-chromosome group of mutant characters of *Drosophila melanogaster,*" *Publ Carneg Inst,* 253:327-339.

Brink, R.A. (1958). "Paramutation at the *R* locus in maize," *Cold Spr Harb Symp,* 23:379-391.

Brink, R.A. (1962). "Phase changing in higher plants and somatic cell heredity," *Quart Rev Biol,* 37:1-22.

Britten, R.J. and Davidson, E.H. (1969). "Gene regulation for higher cells: A theory," *Science,* 165:349-357.

Britten, R.J. and Davidson, E.H. (1971). "Repetitive and nonrepetitive DNA sequences and a speculation on the origins of evolutionary novelty," *Quart Rev Biol,* 46:111-133.

Britten, R.J. and Kohne, D.E. (1968). "Repeated sequences in DNA," *Science,* 161:529.

Britten, R.J. and Kohne, D.E. (1970). "Repeated segments of DNA," *Scient Amer,* 222:24-31.

Brody, E. and Abelson, J. (1985). "The 'spliceosome' yeast premessenger RNA associates with a 40S complex in a splicing-dependent reaction," *Science,* 228:963-967.

Brown, D.D. and David, I.B. (1968). "Specific gene amplification in oocytes," *Science,* 160:272-280.

Brown, D.D. and Weber, C.S. (1968). "Gene linkage by RNA-DNA hybridization II. Arrangement of the redundant gene sequences for 28S and 18S ribosomal RNA," *J Mol Biol,* 34:681-697.

Brown, D.D., Wensink, P.C., and Jorlan, E. (1972). "A comparison of the ribosomal DNA's *Xenopus laevis* and *Xenopus mulleri*: The evolution of tandem genes," *J Mol Biol,* 63:57-73.

Brown, R. (1831). "Observations on the organs and mode of fecundation in *Orchideae* and *Asclepiadeae,*" *Trans Linn Soc London (Bot),* pp. 77-80.

Brown, W.L. and Wilson, E.O. (1956). "Character displacement," *Systemat Zool, 5:49-64.*

Bunn, C.L., Wallace, W.C., and Eisenstadt, J.M. (1974). "Cytoplasmic inheritance of chloramphenicol resistance in mouse tissue culture cells," *Proc Natl Acad Sci USA,* 71:1681-1685.

Burd, C.G. and Dreyfuss, G. (1994). "Conserved structures and diversity of functions of RNA-editing proteins," *Science,* 265:615

Burgess, R.R., Travers, A.A., Dunn, J.J., and Bautz, E.K.F. (1969). "Factors stimulating transcription by RNA polymerase," *Nature,* 221:43-46.

Burnham, C.R. (1962). Discussions in Cytogenetics. Minneapolis: Burgess, p. 375.

Cairns, J. (1963). "The chromosome of *Escherichia coli,*" *Cold Spr Harb Symp,* 28:43-46.

Caldecott, R.S. and Smith, L. (1952). "A study of X-ray-induced chromosomal aberrations in barley," *Cytologia,* 17:224-242.

Caldwell, R.M., Roberts, J.J., and Eyal, Z. (1970). "General resistance (slow rusting) to *Puccinia recondita* in winter wheat and spring wheats," *Phytopathology,* 60:1287-1295.

Callen, H.G. and Lloyd, L. (1960). "Lampbrush chromosomes of crested newts *Triturus Cristatus* (Laurenti)," *Phil Trans Roy Soc B,* 243:135-219.

Callen, H.G. and Tomlin, S.G. (1950). "Experimental studies on amphibian oocyte nuclei I. Investigation of the structure of nuclear membrane by means of electron microscope," *Proc Roy Soc B,* 137:367-378.

Cameron, J.R., Loh, E.Y., and Davis, R.W. (1979). "Evidence for transposition of dispersed repetitive DNA families in yeast," *Cell,* 16:739-751.

Cannon,W.B. (1929). "Organization for physiological homeostasis," *Phys Rev,* 9:399-412.

Capecchi, M.R. (1967). "Polypeptide chain termination *in vitro*: Isolation of a release factor," *Proc Natl Acad Sci USA,* 58:1144-1151.

Capecchi, M.R. (1994). "Targeted gene replacement," *Scient Am,* 270:34-41.

Capecchi, M.R. and Gussin, G.N. (1965). "Suppression *in vitro*: Identification of a serine-sRNA as a 'nonsense' suppressor," *Science,* 149:417-420.

Carle, G.F. and Olson, M. (1985). "An electron karyotype for yeast," *Proc Natl Acad Sci USA,* 82:3756-3760.

Carothers, E.E. (1917). "The segregation and recombination of homologous chromosomes," *J Morph,* 28:445-462.

Castle, W. (1905). "Recent discoveries in heredity and their bearing on animal breeding," *Pop Sci Mon,* 66:193-207.

Castle, W. (1905). "Yellow mice and gametic purity," *Science,* 24:275-277.

Cavalier-Smith, T. (1978). "Nuclear volume control by nucleoskeletal DNA, selection for cell volume and cell growth rate, and the solution of the DNA c-value paradox," *J Cell Sci,* 34:247

Cazenave, C. and Hélène, C. (1991). "Antisense oligonucleotides," In *Antisense Nucleic Acids and Proteins,* eds. Mol, J.N.M. and vander Krol, R.R. New York: Marcel Dekker, Inc., pp. 47-53.

Cech, T.R. (1985). "Self-splicing RNA: Implications for evolution," *Int Rev Cytol,* 93:3-22.

Cech, T.R., Zaug, A. J., and Grabowski, P.J. (1981). "*In vitro* splicing of the ribosomal RNA precursor of Tetrahymena: Involvement of a guanosine nucleotide in the excision of the intervening sequence," *Cell,* 27:487-496.

Charpak, M. and Dedonder, R. (1965). "Production d 'un 'facteur de compéténce' soluble par *Bacillus subtilis* Marburg ind-168," *C R Hebd Acad Sci Paris,* 260:5638-5640.

Chatton, E. (1925). "*Panoporella perplexa,*" *Ann Sci Natur Zool,* 8:5.

Childs, G., Maxon, R., Cohn, R.H., and Kedes, L. (1981). "Orphons: dispersed genetic elements derived from tandem repetitive genes of eucaryotes," *Cell,* 23:651-663.

Chu, E.H.Y., Thuline, H.C., and Norby, D.E. (1964). "Triploid-diploid chimerism in a male tortoiseshell cat," *Cytogenetics,* 3:1-12.

Chun, E.H.L., Vaughan, M.H., and Rech, A. (1963). "The isolation and characterization of DNA associated with chloroplast preparations," *J Mol Biol,* 7:130-141.

Church, G.M., Stonimski, P.P., and Gilbert, W. (1979). "Pleiotropic mutations within two yeast mitochondrial cytochrome genes block mRNA processing," *Cell,* 18:1209-1215.

Cohen, S.N. (1976). "Transposable genetic elements and plasmid evoluton," *Nature,* 263:731-738.
Cole-Turner, R. (1995). "Religion and gene patenting," *Science,* 270:52.
Collins, J. and Hohn, B. (1978). "Cosmids: A type of plasmid gene-cloning vector that is packaged *in vitro* in bacteriophage λ heads," *Proc Natl Acad Sci USA,* 75:4242
Contente, S. and Dubnau, D. (1979). "Marker rescue transformation by linear plasmid DNA in *Bacilus subtilis,*" *Plasmid,* 2:555-571.
Crick, F.H.C. (1958). "On protein synthesis," *Symp Soc Exp Biol,* 12:138-161.
Crick, F.H.C. (1963). "The recent excitement in coding problem," *Progr Nucleic Acid Res,* 1:163-217.
Crick, F.H.C. (1966). "Codon-anticodon pairing," *J Mol Biol,* 19:548-555.
Crick, F.H.C. (1976). "Linking number and nucleosomes," *Proc Natl Acad Sci USA,* 73:2639-2643.
Crick, F.H.C. and Klug, A. (1975). "Kinky helix," *Nature,* 255:530-533.
Crick, F.H.C., Barnett, L., Brenner, S., and Watts-Tobin, R.J. (1961). "General nature of the genetic code for proteins," *Nature,* 192:1227-1232.
Crouse, H.V. (1960). "The controlling element in sex chromosome," *Genetics,* 45:1429-1443.
Dangeard, P.A. (1919). "Sur le distinction du chondriome des auteurs en vacuome, plastidome at spherome," *C R Acad Sci Paris,* 169:1005-1010.
Dangeard, P.A. (1920). "Plastidome, vacuome et spherome dans *Sellaginella kraccssiana,*" *C R Acad Sci Paris,* 170:301-306.
Danna, K. and Nathans, D. (1971). "Specific cleavage of Simian virus 40 DNA by restriction endonuclease of Hemophilus influenza," *Proc Natl Acad Sci USA,* 68:2913.
Darlington, C.D. (1928). "Studies in Prunus, I and II," *J Genet,* 19:213-256.
Darlington, C.D. (1929a). "Chromosome behaviour and structural hybridity in the *Tradescantiae,*" *J Genet,* 21:207-286.
Darlington, C.D. (1929b). "The significance of chromosome behaviour in polyploids for the theory of meiosis," *John Innes Hort Inst Conf on Polyploidy,* p. 42.
Darlington, C.D. (1931). "Chiasma formation and chromosome pairing in *Fratillaria,*" *Proc Second Int Rot Congr,* p. 189.
Darlington, C.D. (1932a). *Recent Advances in Cytology.* London: Churchill.
Darlington, C.D. (1932b). "The control of the chromosome by the genotype and its bearing on some evolutionary problems," *Am Nat,* 66:25-51.
Darlington, C.D. (1937a). "What is a hybrid?" *J Hered,* 28:308-310.
Darlington, C.D. (1937b). *Recent Advances in Cytology,* second edition. London: Churchill.
Darlington, C.D. (1939a). "Misdivision and the genetic of the centromere," *J Genet,* 37:341-364.
Darlington, C.D. (1939b). *The Evolution of Genetic Systems.* Cambridge: Cambridge Univiversity Press.
Darlington, C.D. (1940). "The origin of isochromosomes," *J Genet,* 39:351-361.

Darlington, C.D. and Mather, K. (1949). *The Elements of Genetics.* London: Allen and Unwin.

Darwin, C. (1859). "On the tendency of species to form varieties and on the perpetuation of varities and species by natural means of selection," *Journ Proc Linnean Soc Zoology,* 3(No. 9):45-62.

Darwin, C. (1876). *The Effects of Cross and Self-Fertilization in the Vegetable Kingdom.* London: Murray.

Davidson, E.H., Hough, B.R., Amenson, C.S., and Britten, R.J. (1973). "General interspersion of repetitive sequences elements in the DNA of *Xenopus,*" *J Mol Biol,* 77:1-23.

Davies, R.L., Fuhrer-Krusi, S., and Kucherlapati, R.S. (1982). "Modulation of transfected gene expression mediated by changes in chromatin structure," *Cell,* 31:521-529.

Davis, R.W. and Hyman, R.W. (1971). "A study in evolution: The DNA base sequence homology between coliphages T_7 and T_3, *J Mol Biol,* 62:287-301.

de Beer, G.R. (1930). *Embryology and Evolution.* Oxford: Clarendon Press, p. 161.

De Duve, C. (1963). The lysosome concept. In *Ciba Foundation Symposium on Lysosomes,* ed. Reuck, A.V.S. and Cameron, M.P. Boston: Little, Brown, and Company.

De Duve, C., Pressman, B.C., Gianetto, R., Wattiaux, R., and Appelmans, F. (1955). "Tissue fractionation studies 6. Intracellular distribution patterns of enzymes in rat-liver tissue," *Biochem J,* 60:604-617.

Delbrük, M. and Stent, G.S. (1957). "On the mechanism of DNA replication," In *The Chemical Basis of Heredity,* eds. McElory, W.D. and Glass, B. Baltimore: John Hopkins Press.

Demerc, M. (1956). "Terminology and nomenclature," In *Genetic Studies with Bacteria, Publ Carneg Inst Wash,* No. 612.

Dempster, E.R. and Lerner, I.M. (1950). "Heritability of threshold characters," *Genetics,* 35:212-234.

Dermen, H. (1940). "Colchicine polyploidy and technique," *Bot Rev, 6:599-635.*

DeVries, H. (1901). "Die Mutation theorie Veit Leipzig," (Translated into English and reprinted 1909), Chicago: Open Court Publishing Company.

Diener, T.O. (1971). "Potato spindle tuber 'viruses' IV. A replicating, low molecular weight RNA," *Virology,* 45:411-428.

Ding, H-F., Rimsky, S., Batson, S.C., Bustin, M., and Hansen, U. (1994). "Stimulation of RNA polymerase II elongation by chromosomal protein HMG-14," *Science,* 265:796-799.

Dintzis, H.M., Borsook, H., and Vinograd, J. (1958). *Microsomal Particles and Protein Synthesis,* ed. Roberts, R.E. New York: Pergamon.

Dobzhansky, T. (1935). "A critique of the species concept in biology," *Phil Sci,* 2:344-355.

Dobzhansky, T. (1937). *Genetics and Origin of Species* (first edition). New York: Columbia University Press, p. 364.

Dobzhansky, T. (1946). "Genetics of natural populations. XIII. Recombination and variability in populations of *Drosophila pseudoobscura*," *Genetics,* 31:269-290.
Dobzhansky, T. (1950). "Mendelian populations and their evolution," *Am Nat,* 84:401-418.
Dobzhansky, T. (1951). Mendelian populations and their evolution. In *Genetics in the 20th Century,* ed. Dunn, L.C. New York: MacMillan, p. 573.
Doctor, B.P., Apgar, J., and Holley, R.W. (1961). "Fractionation of yeast amino acid-acceptor ribonucleic acids by countercurrent distribution," *J Biol Chem,* 236:1117-1120.
Dover, G. (1982). "Molecular drive: A cohesive mode of species evolution," *Nature,* 299:111-117.
Dover, G. and Coen, E. (1981). "Spring cleaning ribosomal DNA: A model for multigene evolution," *Nature,* 290:731-732.
Dover, G. and Ford-Doolite, F.W. (1980). "Modes of genome evolution," *Nature,* 288:646-647.
Drake, J.W. and Allen, E.F. (1968). "Antimutagenic DNA polymerase of bacteriophage T_4," *Cold Spr Harb Symp,* 33:339-344.
Dressler, D. (1975). "The recent excitement in the DNA growing point problem," *Ann Rev Microbiol,* 29:525-559.
Dreyfuss, G., Phillipson, L., and Mattaj, I .W. (1988). "Ribonucleoprotein particles in cellular processes," *J Cell Biol,* 106:1419-1425.
Dreyfuss, G., Matunis, M.J., Pinol-Roma, S., and Burd, C.G. (1993). "hnRNP proteins and the biogenesis of mRNA," *Ann Rev Biochem,* 62:289-321.
Du Praw, E.J. (1970). *DNA and Chromosomes.* New York: Rinehard and Winston.
Du Rietz, E. (1930). "The fundamental units of botanical taxonomy," *Svensk Bot Tidskr,* 24:333-428.
Dujardin, F. (1841). *Histoire naturelle des Zoophytes.* Paris: Roret.
Echols, H. (1963). "Properties of F′ strains of *Escherichia coli* superinfected with F-lactose and F-galactose episomes," *J Bact,* 85:262-273.
Echols, H. and Goodman, F. (1991). "Fidelity mechanism in DNA replication," *Ann Rev Biochem,* 60:477.
Edgar, R.S. and Lielausis, I. (1964). "Temperature-sensitive mutants of bacteriophage T4D: Their isolation and genetic characterization," *Genetics,* 49:649-662.
Edmond, M. and Abrams, R. (1960). "Polynucleotide biosynthesis: Formation of a sequence of adenylate units from adenosine triphosphate by an enzyme from thymus nuclei," *J Biol Chem,* 235:1142-1149.
Ege, T. and Ringertz, N.R. (1974). "Preparation of microcells by enucleation of micronuclease cells," *Exp Cell Res,* 87:378-382.
Eghlom, P.E., Buchardt, O., Christensen, L., Behrens, C., Freler, S.M., Driver, D.A., Berg, R.H., Kein, S.K., Norden, B., and Nielsen, P.E. (1993). "PNA hybridizes to complementary oligonucleotides obeying the Watson-Crick hydrogen-bonding rules," *Nature,* 365:566-568.
Eigen, M. (1971). "Self-organization of the matter and evolution of biological macromolecules," *Naturwiss,* 58:465-523.

Ellis, J. (1982). "Promiscuous DNA-chloroplast genes inside plant mitochondria," *Nature,* 299:678-679.

Engler, A. and Prantl, K. (1897). *Die Natürlichen Pflanzenfamilien.* Leipzig: Engelmann.

Epstein, W. and Beckwith, J.R. (1968). "Regulation of gene expression," *Ann Rev Biochem,* 37:411-436.

Farmer, J.B. and Moore, J.E.S. (1905). "On the meiotic phase (reduction divisions) in animals and plants," *Quart J Micr Sci,* 48:489-507.

Fenner, F. (1976). "The classification and nomenclature of viruses: Summary of results of Meeting of the International Committee on Taxonomy of Viruses in Madrid, September 1975," *Virology,* 71:371-378.

Finch, J.T. and Klug, A. (1976). "Solenoidal model for superstructure in chromatin," *Proc Natl Acad Sci USA,* 73:1897-1901.

Finch, J.T., Lutter, L.C., Rhodes, E.D., Brown, R.S., Rushton, B., Levitt, M., and Klug, A. (1977). "Structures of nucleosome core particles of chromatin," *Nature,* 269:29-36.

Finnegan, P. and Brown, G. (1986). "Autonomously replicating RNA in mitochondria of maize plants with S-type cytoplasm," *Proc Natl Acad Sci USA,* 83:5175-5179.

Fischer, E. (1939). "Versch einer Phänogenetik der normalen körperlichen Eigenschaften des Menschen," *Z Indukt Abstamm-u VererbLehre,* 76:47-52.

Fisher, R.A. (1928). "The possible modification of the response of the wild type to recurrent mutations," *Am Nat,* 62:115-126.

Flavell, R.B., Rimpau, J., and Smith, D.B. (1977). "Repeated sequence DNA relationships in four cereal genomes," *Chromosoma,* 63:205-222.

Flemming, W. (1882). *Zellsubstanz, Kern and Zelteilung.* Leipzig: Vogel.

Flock, J.I. (1983). "Cosduction: Transduction of *Bacillus subtilis* with phage ϕ105 cosplasmid," *Mol Gen Genet,* 189:304-308.

Flor, H.H. (1956). "Complementary genetic systems in flax rust," *Adv Genet,* 8:29-54.

Focke, W.O. (1881). *Die Pflanzenmischlinge, ein Beitrag zur Biologie der Gewächse.* Berlin: Borntraeger.

Fol, H. (1877). "Sur le commancement de lhenogenie chez divers animaux," *Arch Sci Nat et Phys Geneve,* 58:79-87.

Földes, J. and Trautner, T.A. (1964). "Infections DNA from a newly isolated *B. subtilis* phage," *Z VererbLebre,* 95:57-63.

Ford, E.B. (1940). Polymorphism and taxonomy. In *New Systematics,* ed. Huxley, J.S. Oxford: Clarendon Press, pp. 493-513.

Fox, S.W. and Dose, K. (1972). *Molecular Evolution and Origin of Life.* San Francisco: W. H. Freeman.

Franklin, N.C. and Luria, S.E. (1961). "Tranduction by bacteriophage P1 and the properties of the *lac* genetic region of *E. coli* and *S. dysenteriae,*" *Virology,* 15:299-311.

Freese, E. (1959). "The difference between spontaneous and base-analogue induced mutations of phage T_4," *Proc Natl Acad Sci USA,* 45:622-633.

Freese, E., Phaese, H.J., and Freese, E.B. (1969). "Chemical DNA alterations causing inactivation and chromosome breaks," *Jap J Genet,* 44 (Suppl 1):94-101.
Fuller, F.B. (1971). "The writhing number of a space curve," *Proc Natl Acad Sci USA,* 68:815-819.
Gall, J.G. and Pardue, M.L. (1969). "Formation and detection of RNA-DNA hybrid molecules in cytology preparations," *Proc Natl Acad Sci USA,* 63:378-383.
Ganoza, M.C. (1966). "Polypeptide chain termination in cell-free extracts of *E. coli,*" *Cold Spr Harb Symp,* 31:273-278.
Gause, G.F. (1934). *The Struggle for Existence.* Baltimore: Williams and Wilkins, p. 163.
Gehring, W.J. (1987). "Homeoboxes in the study of development," *Science,* 236:1245-1252.
Geitler, L. (1939). "Die entstehung der polypleiden somakerne der heleropteren durch chromosomenteilung ahue kernteilung," *Chromosoma,* 1:1-22.
Gellert, M., Mizunchi, K., O'Dea, M.H., and Nash, H.A. (1976). "DNA gyrase: An enzyme that introduces superhelical turns into DNA," *Proc Natl Acad Sci USA,* 73:3872-3876.
Georgiev, G.P. and Mantieva, V.L. (1962). "The isolation of DNA-like RNA and ribosomal RNA from the nucleo-chromosomal apparatus of mammalian cells," *Biochem Biophys Acta* 61:153-54.
Gesterland, R.F., Weiss, R.B., and Atkins, J.F. (1992). "Recoding: Reprogrammed genetic decoding," *Science,* 257:1640-1641.
Gierer, A. (1977). "Physical aspects of tissue evagination and biological form," *Q Rev Biophys,* 10:529-593.
Gilbert, W. (1978) . "Why genes in pieces?" *Nature,* 271:501.
Gilbert, W. and Dressler, D. (1968). "DNA replication: The rolling circle model," *Cold Spr Harb Symp,* 33:473-484.
Gilmore, R.A. (1967). "Supersuppressors in *Saccharmoyces cerevisiae,*" *Genetics,* 56:641-658.
Gilmour, J.S.L. and Gregor, J.W. (1939). "Demes: A suggested new terminology," *Nature,* 144:333.
Goffeau, A. (1995) . "Life with 482 genes," *Science,* 270:445-446.
Goffeau, A., Barrell, B.G., Eussy, H., Davis, R.W., Dujon, B., Feldmann, H., Galibert, F., Hoheisel, J.D., Jaeq, C., Johnston, M., Lovis, E.J., Mewes, H.W., Murakami, Y., Phillipsen, P., Tettelin, H., and Oliver, S.G. (1996). "Life with 6000 genes," *Science,* 274:546-567.
Goldberg, M. (1979). *Sequence analysis of Drosophila histone genes,* PhD Dissertation, Stanford University, p. 173.
Goldman, R.D., Lazarides, E., Pollack, R., and Webster, K. (1975). "The distribution of actin in non-muscle cells," *Exp Cell Res,* 90:333-344.
Goldschmidt, R. (1915). "Vorläufige Mitteilung über weitere Versuche zw Verebung und Besttimmung des Geschlechts," *Biol Zbl,* 35:565-570.
Goldschmidt, R. (1917). "Intersexuality and the endocrine aspect of sex," *Endocrinology,* 1:433-456.

Goldschmidt, R. (1945). "The structure of podoptera, a homoeotic mutant of *Drosophila melanogaster,*" *J Morph,* 77:71-103.

Goodman, S.D. and Nash, H.A. (1989). "Functional replacement of a protein-induced bend in a DNA recombination site," *Nature,* 341:251-254.

Gordon, J.W. and Ruddle, F.H. (1981). "Integration and stable germ line transmission of genes injected into mouse pronuclei," *Science,* 214:1244-1246.

Gradbowski, P.J., Seiler, S.R., and Sharp, P.A. (1985). "A multicomponent complex is involved in the splicing of messenger RNA precursors," *Cell,* 42:345-353.

Grant, V. (1966). "The selective origin of incompatability barriers in the plant genus *Gilia,*" *Am Nat,* 100:99-118.

Gregoire, V. (1905). "Les resultats acqius sur le cinesis de maturation dans les deux regnes. I," *Cellule,* 21:297-315.

Griffing, B. (1956). "Concept of general and specific combining ability in relation to diallel crossing system," *Aust J Biol Sci,* 9:463-493.

Griffith, J.D. (1975). "Chromatin structure: Deduced from a minochromosome," *Science,* 187:1202-1203.

Grinnell, J. (1917). "The niche-relationships of the California thrasher," *Auk,* 34:427-433.

Grodzickev, T., Williams, J., Sharp, P., and Sambrook, J. (1974). "Physical mapping of temperature-sensitive mutations of adenoviruses," *Cold Spr Harb Symp,* 39:439-446.

Gruss, P., Dhar, R., and Khoury, G. (1981). "Simian virus 40 tandem repeated sequences as an element of the early promoter," *Proc Natl Acad Sci USA,* 78:943-947.

Guarrente, L. (1984). "Yeast promoters: Positive and negative elements," *Cell,* 36:799-800.

Haeckel, E. (1866). *Generelle Morphologie der Organismen.* Berlin: Reimer.

Haeckel, E. (1894). *Systematische Phylogenie.* Berlin: Reimer.

Haecker, V. (1892). "Die heterotypische Kernteilung in Zyklus der generativen zellen," *Ber Naturf Ges Freiburg In Br,* 6:160-165.

Haecker, V. (1897). "Über weikere Übereinstimmungen zwischen den Fortpflanzagsvorgängen der Tiere und Planzen. Die Keim-und Mutterzellen," *Biol Zbl,* 17:689-694.

Hajduk, S.L., Harris, H.E., and Polland, V.W. (1993). "RNA editing in kinetoplastid mitochondria," *FASEB,* 7:54-63.

Haldane, J.B.S. (1919). "The combination of linkage values and the calculation of distances between the loci of linked factors,"*J Genet,* 8:299-309.

Haldane, J.B.S. (1942). *New Paths in Genetics.* New York: Harper and Brothers.

Haldane, J.B.S. (1957). "The cost of natural selection," *J Genet,* 55:511-524.

Hamilton, W.D. (1964). "Genetical evolution of social behavior, I and II," *J Theort Biol,* 7:1-16, 17-52.

Hanahan, D., Lane, D., Lipsich, L., Wigler, M., and Botchan, M. (1980). "Characteristics of an SV40-plasmid recombinant and its movement into and out of the genome of a murine cell," *Cell,* 21:127-139.

Hardy, G.H. (1908). "Mendelian proportions in a mixed population," *Science,* 28:49-50.

Harris, H. (1968). "Molecular evolution: The neutralist-selectionist controversy," *Fed Proc,* 35:2079-2082.

Harris, P. and James, T. (1952) "Electron microscope study of the nuclear membrane of *Amoeba foroteusin* thin section," *Experientia,* 8:384-385.

Hartwell, L.H. and Kastan, M.B. (1994). "Cell cycle control and cancer," *Science,* 266:1821-1827.

Harvey, E.B. (1917). "A review of the chromosome number in the metazoa I," *J Morphol,* 28:1-63.

Haselkorn, R. and Fried, V.A. (1964). "Cell-free protein synthesis: The nature of the active complex," *Proc Natl Acad Sci USA,* 51:308-315.

Hawthrone, D.C. and Mortimer, R.K. (1963). "Super-suppressors in yeast," *Genetics,* 48:617-620.

Hayashi, M. Hayashi, M.N., and Spiegelman, S. (1964). "DNA circularity and the mechanism of strand selection in the genetic messages," *Proc Natl Acad Sci USA,* 51:351-359.

Hedges, R.W. and Jacob, A.E. (1974). "Transposition of Amplicillin resistance from RP4 to other replicons," *Mol Gen Genet,* 132:31-40.

Heffron, F., So, M., and McCarthy, J. (1978). "*In vitro* mutagenesis of a circular DNA molecule by using synthetic restriction sites," *Proc Natl Acad Sci USA,* 75:6012-6016.

Heidenhain, M. (1894). "Neue Untersuchungen ümber die Zentralkörper und ihre Beziehungen zum Kern-und-Zellenprotoplasma," *Arch Mikr Anat,* 43:423-424.

Heitz, E. (1928). "Das Heterochromatin der mosse I," *Jahrb Wiss Bot,* 69:762-818.

Heitz, E. (1929). "Heterochromatin chromocentren, chromeren," *Ber Detsch Dot Ges,* 47:274-284.

Heitz, E. (1931). "Beitrag zur Lehre von der Entstehung der karyokinetischen spindel," *Planta,* 12:495-505.

Hentschel, C.C. (1982). "Homocopolymer sequences in the spacer of a sea urchin histone gene repeat are sensitive to S1 nuclease," *Nature,* 295:714-716.

Hermann, H. (1891). "Beitrag zur Lehre voon der Entstehung der karyokinetischen spindel," *Arch Mikr Anat,* 52:314-337.

Hirsch, H.J., Starlinger, P., and Brachet, P. (1972). "Two kinds of insertions in bacterial gene," *Mol Gen Genet,* 119:191-206.

Hoagland, M.B. (1959). "The present status of the adaptor hypothesis," *Brookh Symp Biol,* 12:40.

Hofmeister, W. (1851). *Vergleichende Untersuchungen über die Befurchtung der Koniferen.* Leipzig: Engelmann.

Hofstetter, H.A. (1981). "A split promoter for a eukaryotic tRNA gene," *Cell,* 24:573-585.

Hood, L. (1972). "Two genes, one polypeptide chain—fact or fiction?" *Fed Proc,* 31:177-187.

Hood, L., Campbell, J.H., and Elgin, S.C.R. (1975). "The organization, expression and evolution of antibody genes and other multigene families," *Ann Rev Genet,* 9:305-353.

Hooke, R. (1665). Micrographia, or some physiological descriptions on minute bodies by magnifying glasses. Facsimile in *Early Science in Oxford XIII: The Life and Works of Hooke, R.,* ed. Gunther, R.T. Oxford.

Hotta, Y., Tabata, S., and Stern, H. (1984). "Replication and nicking of zygotene DNA sequences: Control by a meiosis-specific protein," *Chromosoma,* 90:243-253.

Hotta, Y., Tabata, S., Stubbs, L., and Stern, H. (1985). "Meiosis-specific transcripts of a DNA component replicated during chromosome pairing: Homology across the phylogenetic spectrum," *Cell,* 40:785-793.

Howard, A. and Pelc, S.R. (1953). "Synthesis of deoxyribonucleic acid in normal and irradiated cells and its relation to chromosome breakage," *Heredity 6,* (Suppl):261-274.

Hull, F.H. (1946). "Regression analysis of corn yield data," (Abstr.) *Genetics,* 31:219.

Hutchinson, C.A., Phillips, S., and Edgell, H.M. (1978). "Mutagenesis at a specific position in a DNA sequence," *J Biol Chem,* 253:6551-6560.

Huxley, J.S. (1939). "Clines, an auxiliary method in taxonomy," *Bijdr Dierk,* 27:491-520.

Huxley, J.S. (1957). "The three types of evolutionary processes," *Nature,* 180: 454-455.

Ilyin, Y.V., Chmeliauskaite, V.G., Ananiev, E.V., and Georgiev, G.P. (1980). "Isolation and characterization of a new family of mobile dispersed genetic elements mdg 3, in *Drosophila melanogaster*," *Chromosoma,* 81:27.

Ishitsuka, H. and Kaji, A. (1970). "Release of tRNA from ribosomal by a factor other than G factor," *Proc Natl Acad Sci USA,* 66:168-173.

Itoh, T. and Tomizawa, J. (1980). "Formation of an RNA primer for initiation of replication of ColE1 DNA by ribonuclease H," *Proc Natl Acad Sci USA,* 77:2450-2454.

Izant, J.G. and Weintraub, H. (1985). "Constitutive and conditional suppression of exogenous and endogenous genes by antisense RNA," *Science,* 229:345-352.

Jackson, E.N. and Yanofsky, C. (1973). "The regulation between the operator and first structural gene of the tryptophan operon of *Escherichia coli* may have a regulatory function," *J Mol Biol,* 76:89-101.

Jackson, E.N. and Yanofsky, C. (1974). "Location of two functions of the phosphoribosyl anthranilate transferase of *Escherichia coli* to distinct regions of the polypeptide chain," *J Bact,* 117:502-508.

Jacob, F. and Brenner, S. (1963). "Sur la regulation de la synthese du DNA chez les bacteries: 1 hypothese du replicon," *C R Acad Sci Paris,* 256:298.

Jacob, F. and Monod, J. (1959). "Genes de structure at genes de regulation dans la biosynthese des proteins," *C R Acad Sci Paris,* 249:1282-1301.

Jacob, F. and Monod, J. (1961a). "Gene regulatory mechanisms in synthesis of proteins," *J Mol Biol,* 3:318-356.

Jacob, F. and Monod, J. (1961b). "On the regulation of gene activity," *Cold Spr Harb Symp,* 26:193-211.
Jacob, F. and Monod, J. (1965). "Deletions fusionnant l′ operon lactose et un operon purine chez *Escherichia coli,*" *J Mol Biol,* 13:704.
Jacob, F. and Wollman, E.L. (1958). "Les episomes, elementes genetiques ajoutes," *C R Acad Sci Paris,* 247:154-156.
Jacob, F., Brenner, S., and Cuzin, F. (1964). "On the regulation of DNA replication in bacteria," *Cold Spr Harb Symp,* 28:329-348.
Jacob, F., Lwoff, A., Siminoviteh, L., and Wollman, E.L. (1953). "Definition de quelques termes retalifs a la lysogenie," *Ann Inst Pasteur,* 84:222.
Jacob, F., Perrin, D., Sanchez, C., and Monod, J. (1960). "L′ óperon: groupe de génes á expression coardonée pare un opérateur," *C R Acad Sci Paris,* 250:1727-1729.
Jacq, C., Miller, J.R., and Brownlee, G.G. (1977). "A pseudogene structure in 5sDNA of *Xenopus laevis,*" *Cell,* 13:109-120.
Janssens, F.A. (1909). "Spermatogenase dens les Batraciens La theories de la chiasmatype, Nouvelles, interpretation des cineses de maturation," *Cellule,* 25:387-411.
Jeffreys, A.J. and Flavell, R.A. (1977). "The rabbit β-globin gene contains a large insert in the coding sequence," *Cell,* 12:1097-1108.
Jeffreys, A.J., Wilson, V., and Thein, S.L. (1985). "Hypervariable 'minisatellite' regions in human DNA," *Nature,* 314:67-73.
Johannsen, W. (1903). *Über Erblichkeit in Populationene und reinen Linien.* Jena, Germany: Fischer.
Johannsen, W. (1909). *Elemente der exakten Erblichkeitslehre.* Jena, Germany: Fischer.
Johannsen, W. (1926). *Elemente der exakten Erblichkeitslehre* (3. Aufl), Jena: Fischer.
John, H.A., Birnsteil, M.L., and Jones, X.W. (1969). "RNA-DNA hybrids at cytological level," *Nature,* 223:582-587.
Jose, E.G., Pastor, J.L., and SanMillan, F.M. (1994). "The smallest known gene," *Nature,* 369:281.
Kada, T., Inoue, T., and Mamiki, M. (1981). In *Environmental Desmutagens and Antimutagens,* ed. Kletowski, E.J. Praeger, pp. 134.
Kada, T., Kaneko, K., Matsuzaki, S., Matsuzaki, T., and Hara, Y. (1985). "Detection and chemical identification of bioantimutagens. A case of the green tea-factor," *Mutation Research,* 150:127-132.
Kahn, M.E. and Helinski, D.R. (1978). "Construction of a novel plasmid-phage hybrid: Use of the hybrid to demonstrate ColE1 DNA replication *in vivo* in the absence of a ColE1 specific protein," *Proc Natl Acad Sci USA,* 75:2200-2204.
Kasai, T. (1974). "Regulation of the expression of the histidine operon in *Salmonella typhimurium,*" *Nature,* 249:523-527.
Katayama, Y. (1934). "Haploid formation by X-rays in *Triticum monococcum,*" *Cytologia*, 5:235-247.
Kelly, R.B., Cozzarelli, N.R., Drutscher, M.P., Lehman, I.R., and Korenberg, A. (1970). "Enzymatic synthesis of deoxyribonucleic acid. XXXII. Replication

of duplexdeoxyribonucleic acid by polymerase activity at a single strand break," *J Biol Chem,* 245:39.

Kelner, A. (1949). "Photoreactivation of ultraviolet-irradiated *Escherichia coli* with special reference to the dose-reduction principle and to ultraviolet-induced mutation," *J Bact,* 58:511-522.

Khush, G.S. and Rick, C.M. (1967). "Novel compensating trisomics of the tomato: Cytogenetics, monosomic analysis, and other applications," *Genetics,* 56:297-307.

Kihara, H. and Ono, T. (1926). "Chromosomenzahlen und systematische Gruppierung der Rremex-Artan," *Z Zellforsch,* 4:475-482.

Kimber, F. and Sears, E.R. (1968). "Nomenclature for the description of aneuploids in the Triticinae," In *Proc 3rd Int Wheat Symp,* Canberra: Australian Academy of Sciences, pp. 468-473.

Kimura, M. (1956). "A model of a genetic system which leads to closer linkage by natural selection," *Evolution,* 10:278-287

King, J.L. and Jukes, T.H. (1969). "Non-Darwinian evolution," *Science,* 164:788-798.

King, J.P. and Ohta, T. (1975). "Polyallelic mutational equilibria," *Genetics,* 79:681-691.

King, R.C. (1970). "The meiotic behavior of the *Drosophila* oocyte," *Int Rev Cytol,* 28:125-168.

Kit, S. (1961). "Equilibrium sedimentation in density gradients of DNA preparations from animal tissues," *J Mol Biol,* 3:711-716.

Korenberg, A., Lehman, I.R., and Schims, E.S. (1956). "Polydeoxynucleotide synthesis by enzymes from *Escherichia coli,*" *Federation Proc,* 15:291-292.

Kosambi, D.D. (1944). "Crossing-over and recombination frequency," *Ann Eugenics,* 12:172-195.

Kossel, A. (1884). "Histones," *Z Physiol Chemie,* 8:511-515.

Kuhn, B., Abdel-Monem, M., and Hoffmann-Berling, H. (1978). "DNA helicases," *Cold Spr Harb Symp,* 43:63-76.

Kumar, C.C. and Novick, R.P. (1985). "Plasmid pT181 replication is regulated by two counter transcripts," *Proc Natl Acad Sci USA,* 82:638-642.

Kurland, C.G. (1960). "Molecular characterization of ribonucleic acid from *Escherichia coli* ribomes," *J Mol Biol,* 2:83-91.

Lagerkvist, U. (1978). "Two out of three: An alternative method of codon reading," *Proc Natl Acad Sci USA,* 75:1759-1762.

Lamb, C.J. (1994). "Plant disease resistance genes in signal perception and transduction," *Cell,* 76:419-422.

Lander, E.S., Green, P., Abrahmson, J., Barlow, A., Daly, M.J., Lincoln, S.E., and Newburg, L. (1987). "MAPMAKER: An interactive computer package for constructing primary genetic linkage maps of experimental and natural populations," *Genomics,* 1:174-181.

Landschulz, H.W., Johnson, P.F., and McKnight, S.L. (1988). "The leucine zipper: A hypothetical structure common to a new class of DNA binding proteins," *Science,* 240:1759.

Langlet, O.F. (1927). "Beiträge zur Cytologie der Ranunculaceen," *Svensk Bot Tidskr,* 21:1-15.
Lark, K.G. (1972). "Genetic control over the initiation of the synthesis of short deoxynucleotide chains in *E. coli,*" *Nature New Biol,* 240:237-240.
Larkin, P.J. and Scowcroft, W.R. (1981). "Somaclonal variation—a novel source of variability from cell cultures for plant improvement," *Theor Appl Genet,* 60:197-214.
Latt, S.A. (1973). "Microfluorometric detection of deoxyribonucleic acid replication in human metaphase chromosomes," *Proc Natl Acad Sci USA,* 70:3395-3399.
Lawrence, J.B. (1990). In *Genome Analysis Vol 1: Genetic and Physical Mapping,* ed. Davies, H.E. and Tilghman, S.M. pp. 1.
Lazowska, J., Jacq, C., and Slonimski, P.P. (1980). "Sequence of introns and flanking exons in wild-type and *box3* mutants of cytochrome reveals an interlaced splicing protein coded by an intron," *Cell,* 22:333-348.
Lederberg, J. (1952). "Cell genetics and hereditary symbiosis," *Physio Rev,* 32:403-430.
Lederberg, J. (1955). "Recombination mechanisms in bacteria," *J Cell Comp Physiol,* 45 (Suppl II):75-107.
Lederberg, J., Cavelli, L., and Lederberg, E.M. (1952). "Sex compatability in *Escherichia coli,*" *Genetics,* 37:720.
Lee, C.S., Davis, R.W., and Davidson, N. (1970). "A physical study by electron microscopy of the terminally repetitious, circularly premuted DNA from the coliphage particles of *Escherichia coli,* 15," *J Mol Biol,* 48:1-22.
Lee, D.H., Granja, J.R., Martinez, J.A., Severin, K., and Ghadori, M.R. (1996). "A self-replication peptide," *Nature,* 382:525-528.
Lee-Huang, S. and Ochoa, S. (1972). "Specific inhibitors of MS2 and late T4 RNA translocation in *E. coli,*" *Biochem Biophys Res Com,* 49:371-376.
Leib, M. (1966). "Studies on heat-inducible lambda bacteriophage I. Order of genetic sites and properties of mutant prophages," *J Mol Biol,* 16:149-164.
Lerner, I.M. (1954). *Genetics Homeostasis.* Edinburgh, Scotland: Oliver and Boyd.
Levan, A. and Hauschka, T.S. (1953). "Endomitotic reduplication mechanisms in ascites tumors of the mouse," *J Nat Cancer Inst,* 14:1-43.
Levene, P.A. and Bass, L.W. (1931). *Nucleic Acids.* New York: Chemical Catalog Co., p. 415.
Levitsky, G.A. (1924). *Materielle Grundlagen der Vererbung* (Russian). Staatsverlag: Kiew.
Lewis, D. (1954). "Comparative incompatability in angiosperms and fungi," *Adv Genet,* 6:235-285.
Lewis, E.B. (1948). "Pseudoallelism in *Drosophila melanogaster,*" *Genetics,* 33:113.
Lewis, E.B. (1951). "Pseudoallelism and gene evolution," *Cold Spr Harb Symp,* 16:159.
Lewis, H. (1966) "Speciation in flowering plants," *Science,* 152:167-172.

Li, H.W., Pao, W.K., and Li, C.H. (1945). "Desynapsis in the common wheat," *Am J Bot,* 32:92-101.

Lindergren, C.C. (1953). "Gene conversion in *Saccharomyces,*" *Jour Genet,* 51:625-637.

Linn, S. and Arber, W. (1968). "Host specificity of DNA produced by *Escherichia coli,* X. *In vitro* restriction of phage FD replication form," *Proc Natl Acad Sci USA,* 59:1300-1306.

Linn, S., Kuhnlein, U., and Deutsch, W.A. (1979). In *DNA Repair Mechanisms,* ed. Hanawalt, P.C., Friedberg. E.C., and Fox, C.F. New York: Academic Press.

Litt, M. and Lutty, J.A. (1989). "Microsatellites," *Am J Hum Genet,* 44:397-401.

Ljungdahl, H. (1924). "Ümber die Herkunft der in der Meiosis konjugierenden chromosomen lei Papeuer-Hybriden," *Svensk Bot Tidskr,* 18:279-291.

Longley, A.E. (1937). "Morphological characters of teosinte chromosomes," *J Agric, Res,* 54:835-862.

Longley, A.E. (1945). "Abnormal segregation during megasporogenesis in maize," *Genetics,* 30:100-105.

Low, R.L., Arai, K., and Korenberg, A. (1981). "Conservation of primosome in successive stages of ϕX174 DNA replication," *Proc Natl Acad Sci USA,* 78:1436-1440.

Lowry, O.H., Rosebrough, J.N., Farr, A.L., and Randell, R.J. (1951). "Protein measurement with the folin phenol reagent," *J Biol Chem,* 193:265-275.

Lucas-Lenard, J. and Lipman, F. (1966). "Separation of three microbial amino acid polymerization factors," *Proc Natl Acad Sci USA,* 55:1562-1574.

Luck, D.J.L. and Reich, E. (1964). "DNA in mitochondria of *Neurospora,*" *Proc Natl Acad Sci USA,* 52:931-938.

Lund-Dahberg, J.E., Lindahl, L., Jaskunas, S.R., Dennis, P.P., and Nomura, M. (1976). "Transfer RNA genes between 16S and 23S rRNA genes in rRNA transcription units of *E. coli,*" *Cell,* 7:165-177.

Lundergardh, H. (1912). "Fixierung, Färbang und Nomenklatur der Kernstruleturen ein Beitrag zur Theorie der zytologischem Methodik," *Arch Mikr Anat,* 80:223-235.

Luria, S.E. and Delbrük, M. (1943). "Interference between bacterial viruses: I. Interference between inactivated bacterial virus and active virus of the same strain and of a different strain," *Arch BioChem,* 1:207-228.

Lush, J.L. (1941). "Methods of measuring the heritability of individual differences among farm animals," *Proc Seventh Intern Congr Genetics 1939 J Genet,* (Suppl Vol.):199

Lush, J.L. (1945). *Animal Breeding in Plants,* third edition. Ames: Iowa State College Press.

Lwoff, A. (1949). "Les organites doués de continuité génétique chez les protistes," *Edition du Centre nat de la Rech Sci Paris,* pp. 7-23.

Lwoff, A. (1953). "Lysogeny," *Bacteriol Rev,* 17:269-337.

Lwoff, A., Anderson, T.F., and Jacob, F. (1959). "Remarques sur les charactéristiques de la particule virale infectreuse," *Ann Inst Pasteur,* 97:281-312.

Magasanik, B. (1961). "Catabolite repression," *Cold Spr Harb Symp,* 26:294-322.

Maire, R. (1902). "Recherches cytologique at taxonomiques sur les Basidiomycetes," *Bull Soc Mycol Fr,* 18:1-32.

Malecot, G. (1948). *Les Mathematique de L'heredite.* Paris: Masson and Cie.

Mantei, N., Boll, W., and Weissman, C. (1979). "Rabbit β-globin mRNA production in mouse cells transformed with cloned rabbit β-globin chromosomal DNA," *Nature,* 281:40-46.

Margaret, E.S., Jeffrey, R.S., and Abelson, J.N. (1994). "The transfer RNA identify problem: A search for rules," *Science,* 263:1411-1447.

Margulis, L. (1970). *Origin of Eukaryotic Cell.* New Haven, CN: Yale University Press.

Markert, C.L. and Møller, F. (1959). "Multiple forms of enzymes: Tissue, ontogenetic, and species specific patterns," *Proc Natl Acad Sci USA,* 45:753-763.

Masiov, D.A. (1994). "Evolution of RNA editing in kinetoplastid protozoa," *Nature,* 368:345-348.

Mather, K. (1933). "The relation between chiasmate and crossing-over in diploid and triploid *Drosophila melanogaster,*" *J Genet,* 27:243-259.

Mather, K. (1941). "Variation and selection of polygenic characters," *J Genet,* 41:159-193.

Mather, K. (1948). "Nucleus and cytoplasm in differentiation," *Symp Soc Exp Biol,* 2:196-216.

Matthey, P.R. (1945). "L' évolution de la formule chromosomiale chex les vertébrés," *Experientia,* 1:50-56 and 78-86.

Maxam, A.M. and Gilbert, W. (1977). "A new method for sequencing DNA," *Proc Natl Acad Sci USA,* 74:560-564.

Mayfield, J.E. and Ellison, J.R. (1975). "The organization of interphase chromatin in Drosophilidae," *Chromosoma,* 52:37-48.

Mayr, E. (1940). "Speciation phenomenon in birds," *Am Nat,* 74:249-278.

Mayr, E. (1942). *Systematics and the Origin of Species.* New York: Columbia University Press.

Mayr, E. (1948). "The bearing of new systematics on general problems. The nature of species," *Adv Genet,* 2:205-237.

Mayr, E. (1963). *Animal Species and Evolution.* Cambridge, MA: Belknap Press of the Harvard University Press.

McBridge, O.W. and Ozer, H.L. (1963). "Transfer of genetic information by purified metaphase chromosomes," *Proc Natl Acad Sci USA,* 70:1258-1262.

McClintock, B. (1934). "The relation of a particular chromosomal element to the development of the nucleoli in *Zea mays,*" *Zeitscher Zellf u Milkr Anat,* 21:294-328.

McClintock, B. (1951). "Chromosome organization and genic expression," *Cold Spr Harb Symp,* 16:13-47.

McClintock, B. (1956). "Controlling elements and the gene," *Cold Spr Harb Symp,* 21:197-216.

McClintock, B. (1978). "Use of transposable genetic elements of maize in gene transfer," *Stadler Symp,* 10:25-47.

McClung, C.E. (1900). "The spermatocyte divisions of the Acrididae," *Kans Univ Quart,* A9:73-84.
McGinnis, W., Garber, R.L., Wirz, J., Kurotwa, A., and Gehring, W.J. (1984). A conserved DNA seqence in homeotic genes of the *Drosophila* antennapedia and bithorax complexes," *Nature,* 308:428-433.
McKnight, S.L. and Kingsburg, R. (1982). "Transcriptional control signals of a eukaryotic protein-coding gene," *Science,* 217:316-324.
McKusick, V.A. and Ruddle, F.H. (1977). "The status of the gene map of the human chromosomes," *Science,* 196:390-405.
Meischer, F. (1869). "On the chemical composition of pus cells." Translated in *Great Experiments in Biology,* eds. Gabriel M.L. and Fogel, S. Englewood Cliffs, NJ: Prentice-Hall, 1955, pp. 233-239.
Mendel, J.G. (1866). "Versuche über Pflanzen-hybriden," *Verh Naturf Ver Brünn,* 4 (Abh).
Meselson, M. and Stahl, F.W. (1958). "The replication of DNA in *Escherichia coli,*" *Proc Natl Acad Sci USA,* 44:671-682.
Meselson, M. and Yuan, R. (1968). "DNA restriction enzyme from *E. coli,*" *Nature,* 217:1110-1114.
Metz, C.W. (1916). "Chromosome studies on the diptera II. The paired association of chromosomes in the diptera, and its significance," *J exp Zool,* 21:213-279.
Miele, E.A., Mulls, D.R., and Kramer, F.R. (1983). "Autocatalytic replication of a recombinant RNA," *J Mol Biol,* 171:281-295.
Minot, C.S. (1908). *The Problems of Age, Growth and Death. A Study of Cytomorphosis.* New York: Putnam's Sons.
Mitsuhashi, M., Miekoogura, A.C., Shinagawa, T., Yano, K., and Hosokawa, T. (1994). "Oligonucleotide probe design—a new approach," *Nature,* 367:759-761.
Mizuno, T., Chou, M., and Inouye, M. (1983). "Regulation of gene expression by a small RNA transcript (mic RNA) in *Escherichia coli* K-12," *Proc Jap Acad Sci,* 59:335-328.
Mohl, H.v. (1846). "Über die Saftbewegungen in Inneren der Zelle," *Bot Ztg,* 4:73-79.
Mohr, O.L. (1923). "Deficiency-Phänomen bei *Drosophila melanogaster,*" *Z indukt Abstamm-u VererbLehre,* 30:279-283.
Monod, J. And Cohen-Bazire, G. (1953). "L 'effect d' inhibition spècifique dans la biosynthèse de la tryptophane-desmase chez *Aerobacter aerogenes,*" *C R Acad Sci Paris,* 236:530-532.
Morgan, T.H. (1910). "Chromosomes and heredity," *Am Nat,* 44:449-496.
Morgan, T.H. (1911). "An attempt to analyse the constitution of the chromosomes on the basis of sex-limited inheritance in *Drosophila,*" *J exp Zool,* 11:365-414.
Morgan, T.H. and Cattell, E. (1912). "Data for the study of sex-linked inheritance in *Drosophila,*" *J exp Zool,* 13:79-101.
Morgan, T.H., Bridges, C.B., Sturtevant, A.H. (1928). "The constitution of the germinal material in relation to heredity," *Yearbook Carneig Inst,* 27:1-28.

Morgan, T.H., Sturtevant, A.H., Muller, H.J., and Bridges, C.B. (1915). *The Mechanism of Mendelian Heredity.* New York: Henry, Holt and Company.
Moses, M.J. (1958). "The relationship between the axial complex of meiotic prophase chromosomes and chromosome pairing in a salamander (*Plethoden cinerens*)," *J Biophys Biochem Cytol,* 4:633-651.
Muir, D.S.F. (1973). Airborne allergens in clinical aspects of inhaled particles. In *Clinical Genetics,* ed. Muir, D.C.F. London: Heinemann Medical Books Ltd., pp. 27-53.
Muller, H.J. (1916). "The mechanism of crossing over," *Am Nat,* 50:284-305.
Muller, H.J. (1932). "Further studies on the nature and causes of gene mutations," *Proc Sixth Int Congr Genet,* 1:213-255.
Muller, H.J. (1934). The effects of X-rays upon the hereditary material. In *Science of Radiology.* London: Bailliere, Tindall and Cox, p. 359.
Muller, H.J. (1940a). "An analysis of the process of structural change in chromosomes of *Drosophila,*" *J Genet,* 40:1-66.
Muller, H.J. (1940b). Bearings of the Drosophila work on systematics. In *The New Systematics.* Oxford: Clarendon Press, p. 185.
Muller, H.J. (1950). "Our load of mutations," *Am J Hum Genet,* 2:111-176.
Muller, H.J. and Altenburg, E. (1919). "The rate of change of hereditary factors in *Drosophila,*" *Proc Soc Exptl Biol NY,* 17:10-14.
Muller, H.J., League, B.B., and Offermann, C.A. (1931). "Effect of dosage changes of sex-linked genes and the compensatory effect of other gene-differences between male and female." *Anat Rec,* 51 (Suppl):110.
Nass, M.K. and Nass, S. (1962). "Fibrous structures within the matrix of developing chick embryos mitochondria," *Exptl Cell Res,* 26:424-437.
Navashin, M.S. (1927). "Über die Veränderung von Zahl und Form der chromosomen infolge der Hybridization," *Zellforsch,* 6:195-233.
Navashin, S.G. (1899). "Neue Beobachtungen über Befruchtung bei *Fritillaria tenella* und *Lilium martagon,*" *Bot Zbl,* 77:62-69.
Navashin, S.G. (1912). "Sur le dimorphisme nucléaire des cellules somatiques der *Galtonia candicans,*" *Bull Acad Sci Imp Petersb Ourg,* 22:373-385.
Niebuhr, E. (1972). "Dicentric and monocentric Robertsonian translocations in man," *Humangenetk,* 16:217-226.
Němec, B. (1910). *Das Problem der Befruchtungsvorgänge und andere zytologische Fragen.* Berlin: Borntraeger.
Nichols, R. (1994). "A universal nucleoside for use at ambiguous sites in DNA primers," *Nature,* 369:492-493.
Nilsson-Ehle, H. (1908). "Einige Ergebnisse Von Krenzungen bei Hafer und Weizen," *Bot Notiser,* pp. 257-294.
Nilsson-Ehle, H. (1911). "Kreuzungunstersuchungen an Hafer und Weizen II," *Lundo Univ Arsskrft NF Avd 7,* No. 6.
Nomura, N. and Ray, D.S. (1980). "Expression of a DNA strand initiation sequence of ColE1 plasmid in single-stranded DNA phage," *Proc Natl Acad Sci USA,* 77:6566-6570.

Nowak, R. (1995), "From genome to proteome: Looking at a cells proteins," *Science,* 270:369-371.
Nudler, E.A., Goldfarb, A., and Kashlev, M. (1994). "Discontinuous mechanism of transcription elongation," *Science,* 265:793-796.
O'Hare, K., Benoist, C., and Breathnack, R. (1981). "Transformation of mouse fibroblasts to methotrexate resistance by a recombinant plasmid expressing a prokaryotic dihydrofolate reductase," *Proc Natl Acd Sci USA,* 78:1527-1531.
Olins, A.L. and Olins, D.E. (1973). "Spheroid chromatin units (v bodies)," *J Cell Biol,* 59:252a.
Oparin, A.I. (1961). *Life: Its Nature, Origin and Development.* Edinburgh, Scotland: Oliver and Boyd.
Oudet, P., Gross-Bellard, M., and Chambon, P. (1975). "Electronmicroscopic and biochemical evidence that chromatin structure is a repeating unit," *Cell,* 4:281.
Padgett, R.A., Konarska, M.M., Grabowski, P.J., Hardy, S.F., and Sharp, P.A. (1984). "Lariat RNA's as intermediates and products in the splicing of messenger RNA precursors," *Science,* 225:898-899.
Paigen, K. and Ganschow, R.S. (1965). "Genetic factors in enzyme realization," *Brookh Symp Biol,* 18:99-115.
Painter, T.S. (1939). "The structure of salivary gland chromosomes," *Am Nat,* 73:315-330.
Painter, T.S. and Muller, H.J. (1929). "Parallel cytology and genetics of induced translocations and deletions in *Drosophila,*" *J Hered,* 20:287-298.
Palmer, J.D. and Shields, C.R. (1984). "Tripartite structure of the *Brassica campestris* mitchondrial genome," *Nature,* 307:437-440.
Pardee, A.B., Jacob, F., and Monod, J. (1959). "The genetic control and cytoplasm expression of 'inducibility' in the synthesis of β-falactosidase by *E. coli,*" *J Mol Biol,* 1:165-178.
Paulson, J.R. and Laemmli, U.K. (1977). "The structure of histone-depleted metaphase chromosomes," *Cell,* 12:817-828.
Perner, E.S. (1952). "Die Vitalfärbung mit Berverinsulfat und inhre physiologische Wirkung auf Zellen höherer Planzen," *Ber dtsch bot Ges,* 65:52-62.
Perucho, M., Hanahan, D., and Wigler, M. (1980). "Genetic and physical linkage of exogenous sequences in transformed cells," *Cell,* 22:309-317.
Philipschenko, J. (1927). *Variabilität und Variation.* Berlin: Borntraeger, p. 101.
Piekarski, G. (1937). "Cytologische Untersuchungen an Paratyyphys-und colilungserscheinungen," *Arch MicroBiol,* 8:428-443.
Pincher, C. (1946). *The Breeding of Farm Animals.* Penguin Handbook, p. 247.
Pinkel, D., Landegent, J., Collins, C., Fuscoe, J., Segraves, R., Lucas, J., and Gray, J. (1988). "Flourescence *in situ* hybridization with human chromosome specific libraries: Detection of trisomy 21 and translocations of chromosome 4," *Proc Natl Acad Sci USA,* 85:9138-9142.
Plate, L. (1910). *Vererbungslehre und Deszendenstheorie, Festschr f Hertwig R II,* Jena, Germany: Fischer. p. 537.
Plate, L. (1913). *Sekktronsprinzip und Probleme der Artbildung.* Berlin: Engelmann.

Pontecarvo, G. (1949). "Auxanographic technique in biochemical *Genetics,*" *J General MicroBiol,* 3:122-126.

Potter, H. and Dressler, D. (1976). "On the mechanism of genetic recombination: Electronic microscopic observation of recombination intermediates," *Proc Natl Acad Sci USA,* 73:3000-3004.

Poulton, E.B. (1903). "What is a species?" *Proc Ent Soc Lond,* 73:26-66.

Price, S. (1956). "Cytological studies in *Saccharum* and allied Genera I. Syncytes in certain clones of *Saccharum and Erianthus,*" *Cytologia*, 21:21-35.

Pring, D.R., Levings, C.S., III. Hu, W.W.L., and Timithy, D.H. (1977). "Unique DNA associated with mitochondria in the '5'-type cytoplasm of male sterile maize," *Proc Natl Acad Sci USA,* 74:2904-2908.

Punnett, R.C. (1905). *Mendelism.* Cambridge: MacMillan, p. 120.

Purkinje, J.E. (1840). Über die Analogien in den Strukturelementen des thierischen und pflanzlichen Organismus. In *Übers Arb Veränder schles Ges vaterl Cultur,* 1839, p. 81.

Radman, M. (1974). In *Molecular and Environmental Aspects of Mutagenesis,* eds. Prakash, L., Sherman, F., Miller, M.W., Lawrence, C.W., and Taber, H.W., Thomas, Springfield III, pp. 128.

Radman, M., Villani, G., Boiteux, S., Defais, M., Laillet-Fouquet, P., and Spaden, S., (1977). In *Origins of Human Cancer,* ed. Watson, J.D., Jiatt, I., and Winston, H. New York: *Cold Spr Harb Lab,* p. 903.

Raff, R.A. and Mahler, H.R. (1972). "The non-symbiotic origin of mitochondria," *Science,* 177:575-582.

Randolph, L.F. (1928). "Chromosome numbers in *Zea mays L.,*" *Cornell Univ Agr Exp Mem,* 117.

Rasmussen, R.E. and Painter, R.B. (1964). "Evidence for repair of ultraviolet damage deoxyribonucleic acid in cultured mammalian cells," *Nature,* 203: 1360-1362.

Ravin, A.W. (1963). "Experimental approaches to the study of bacterial phylogeny," *Am Nat,* 97:307-318.

Razin, A., Hirose, T., Itakura, K., and Riggs, A.D., (1978). "Efficient correction of a mutation by use of chemically synthesized DNA," *Proc Natl Acad Sci USA,* 75:4268-4270.

Renner, O. (1916). "Zur Terminologie des Pflanzlichen Generationswechsels," *Biol Zbl,* 36:337-359.

Rensch, B. (1947). *Neuere Probleme der Abstammungslehre. Die transspesifische Evolution.* Stuttgart: Enke.

Renwick, J.H. (1971). "The mapping of human chromosomes," *Ann Rev Genet,* 35:81-120.

Rickett, H.W. (1958). "So what is a taxon?" *Taxon,* 7:37-39.

Rieger, R.A., Michaelis, A., and Green, M.M. (1991). *Glossary of Genetics.* Berlin: Springer-Verlag, p. 553.

Riley, R. and Chapman, V. (1958). "Genetic control of the cytologically diploid behaviour of hexaploid wheat," *Nature,* 182:713-715.

Ris, H. and Chandler, B.L. (1963). "The ultrastructure of genetic systems in prokaryotes and eukaryotes," *Cold Spr Harb Symp,* 28:1-8.

Roberts, J.W. (1969). "Termination factor for RNA synthesis," *Nature,* 224: 1168-1174.

Roberts, R.B. (1958). *Introduction to Microsomal Particles and Protein Synthesis.* New York: Pergamon.

Robertson, J.D. (1963). "The occurrence of a subunit pattern in the unit membrane of club endings in Mauthner cell synapsis in goldfish brains," *J Cell Biol,* 19:201-221.

Rogers, J. (1983). "Retroposons defined," *Nature,* 301:460.

Roman, H.L. (1956). "Studies of gene mutations in *Saccharomyces,*" *Cold Spr Harb Symp,* 21:175-185.

Roth, C.W. and Nester, E.W. (1971). "Coordinate control of tryptophan histidine and tyrosine enzyme synthesis is *Bacillus subtilis,*" *J Mol Biol,* 62:577-589.

Roth, T.F. and Porter, K.R. (1964). "Yolk protein uptake in the oocyte of the mosquito *Aedes aegypti* L.," *J Cell Biol,* 20:313-332.

Roux, W. (1905). Die Entwicklungsmechanik, ein neuer Zweig der biologischen Wissenschaft. In *Vorträge and Aufsätze über Entwicklungsmechanik der Organismen 1,* ed. Roux, W. Leipzig: Engelmann.

Rowen, L. and Kornberg, A. (1978). "J Primase, the dnaG protein of *Escherichia coli:* An enzyme which starts DNA synthesis," *Biol Chem,* 253:758-764.

Rowlends, T., Baumann, P., and Jackson, S.P. (1994). "The TATA-binding protein: A general transcription factor in eukaryotes and Archaebacteria," *Science,* 264:1326.

Ruddle, F., Riccinti, F.A., McMorris, J., Tischfield, R., Creagan, R., Darlington, G., and Chen, T. (1972). "Somatic cell genetic assignment of peptidase C and the Rh linkage group to chromosome A-1 in man," *Science,* 176:1429-1431.

Rupp, W.D. and Howard-Flanders, P. (1968). "Discontinuities in the DNA synthesized in an excision-defective stain of *Escherichia coli* following ultraviolet irradiation," *J Mol Biol,* 31:291-304.

Russel, S.D. (1980). "Participation of male cytoplasm during gene fusion in an angiosperm *Plumbago zeylanica,*" *Science,* 210:200-201.

Russell, W.L. (1951). "X-ray induced mutations in mice," *Cold Spr Harb Symp,* 16:327-336.

Ryan, F.J. and Lederberg, J. (1946). "Reverse mutation and adaptation in leucineless *Neurospora,*" *Proc Natl Acad Sci USA,* 32:163-173.

Sager, R. (1973). *Cytoplasmic Genes and Organelles.* New York: Academic Press.

Saiki, R.K., Scharf, S., Falooua. F., Mullis, K.B., Horn, G.T., Erlich, H.A., and Arnheim, N. (1985). "Enzymatic amplification of β-genomic sequence and restriction site analysis for diagnosis of sickle cell anemia," *Science,* 230:1350-1354.

Saito, I. and Stark, G.R. (1986). "Charomids: Cosmid vectors, for efficient coloning and mapping of large or small restriction fragments," *Proc Natl Acad Sci USA,* 83:8664-8668.

Sandler, L. and Hiraizumi, Y. (1961). "Meiotic drive in natural populations of *Drosophila melanogaster*. A heritable aging effect on phenomenon of segregation distortion," *Canad J Genet Cytol,* 3:34-46.
Sandler, L. and Novitski, E. (1957). "Meiotic drive as an evolutionary force," *Am Nat,* 91:105-110.
Sanger, F., Nicklen, S., and Coulson, A.R. (1977). "DNA sequencing with chain-terminating inhibitors," *Proc Natl Acad Sci USA,* 74:5463-5467.
Sauer, F., Fondell, J.D., Ohkuma, Y., Reader, R.G., and Jackie, H. (1995). "Control of transcription by Krüppel through interactions with TFIIB and TFIIEβ," *Nature,* 375:563
Sax, K. (1941). "Types and frequencies of chromosomal aberrations induced by X-rays," *Cold Spr Harb Symp,* 9:93-103.
Scaife, J. and Gross, J.D. (1962). "Inhibition of multiplication of F-lac factor in Hfr cells of *Escherichia coli* K-12," *Biochem Biophys Res Comm,* 7:403-407.
Schachat, F.H. and Hogness, D.S. (1973). "Repetitive sequences in isolated Thomas circles from *Drosophila melanogaster*," *Cold Spr Harb Symp,* 38:371-381.
Scherrer, K. and Darnell, J.E. (1962). "Sedimentation characteristics of rapidly labelled RNA from Hela cells," *Biochem Biophys Res Comm*, 7:486-490.
Schimper, A.J.W. (1885). "Untersuchungen über die chlorophyllkörnner und die ihnen homologen Gebilde," *Jahrb Wiss Bot,* 16:1.
Schindler, D. and Echols, H. (1981). "Retroregulation of the int gene of bacteriophage λ: Control of translation completion," *Pro Natl Acad Sci USA,* 78:4475-4479.
Schleicher, W. (1878). "Die Knorpel-Zellteilung,' *Arch Mikr Anat,* 16:248-257.
Schleiden, M.J. and Schwann, T. (1839). In *Great Experiments in Biology,* eds. Gabriel, M. and Fogel, S. Englewood Cliffs, NJ: Prentice-Hall, p. 12.
Schlesinger, D.H., Goldstein, G., and Niall, H.D. (1975). "The complete amino acid sequences of ubiquitin, an adenylate cyclase stimulating polypeptide probably universal in living cells," *Biochemistry,* 14:2214-2218.
Schrader, F. (1936). "The kinetochore or spindle fiber locus in *Amphiuma tridactylus,*" *Biol Bull,* 70:484-498.
Schultz, J. (1936). "Variegation in *Drosophila* and the inert chromosome regions," *Proc Natl Acad Sci USA,* 22:27-33.
Schwann, T. (1838). "Miksoskopische Untersuchungen über die Übereinstimmug in Struktw und dem Wachstum der Thiere und Pflanzen," *Ostwalds Klassiker,* Nr 176., Leipzig: Englemann, p. 72.
Sears, E.R. (1953). "Nullisomic analysis in wheat," *Amer Naturalist,* 87:245-252.
Sears, E.R. (1972). "Chromosome engineering in wheat," *Stadler Genet Symposium,* Columbia: University of Missouri, 4:23-28.
Seeberg, E. and Steinum, A.L. (1982). "Purification and properties of the uvr protein from *Escharichia coli,*" *Proc Natl Acad Sci USA,* 79:988-992.
Seifert, H.S., Chen, E.Y., So, M., and Heffron, F. (1986). "Shuttle mutagenesis: A method of transposon mutagenesis for *Saccharomyces cerevisiae,*" *Proc Natl Acad Sci USA,* 83:735-739.

Seifritz, W. (1928). "Der physkalischen Eigenschaften des Protoplasmas," *Arch Exp Zellforsch,* 6:341-359.

Seiler, J. and Haniel, C.B. (1921). "Das verschiedene Verhalten der chromosomen in Eireifung und Samenreifung von *Lymantria monacha,*" *Z Indukt Abstamm-u VererbLehre,* 27:81-103.

Shibata, T., DasGupta, C., Cunningham, P., and Radding, C.M. (1979). "Purified *Escherichia coli* recA protein catalyzes homologous pairing of superficial DNA and single-stranded fragments," *Proc Natl Acad Sci USA,* 76:1638-1642.

Shine, J. and Dalgarno, L. (1974). "The 3′ terminal sequence of *Escherichia coli* 16S ribosomal RNA: Complementarity to nonsense triplets and ribosome binding sites," *Proc Natl Acad Sci USA,* 71:1342-1346.

Shull, G.H. (1911). "Experiments with maize," *Bot Gaz,* 52:480-531.

Shull, G.H. (1914). "Duplicate genes for capsule form in *Bursa Bursa-pastoris,*" *Z Indukt Abstamm-u VererbLehre,* 12:97-103.

Sigal, N. and Alberts, B. (1972). "Genetic recombination: The nature of a crossed strand-exchange between two homologous DNA molecules," *J Mol Biol,* 71: 789-793.

Silber, R., Malathi, V., and Hurwitz, J. (1972). "Purification and properties of bacteriophage T4-induced RNA ligase," *Proc Natl Acad USA,* 69:3009-3013.

Siliciano, P.G. and Tätchell, K. (1986). "Identification of the DNA sequences controlling the expression of the MATα locus of yeast," *Proc Natl Acad Sci USA,* 83:2320-2324.

Simpson, G.G. (1944). *Tempo and Mode in Evolution.* New York: Columbia University Press.

Simpson, L. (1972). "The kinetoplast of Hemoflagellates," *Int Rev Cytol,* 32: 139-207.

Singer, M.F. (1982). "SINEs and LINEs: Highly repeated short and long intersp-ersed sequences in mammalian genomes," *Cell,* 28:433-434.

Skonju, S., Bogenhagen, D.F., and Brown, D.D. (1980). "A control region in the center of the 5S gene directs specific initiation of transcription. II. The 3′ border region," *Cell,* 19:27-35.

Slautterback, D.B. (1963). "Cytoplasmic microtubules I. Hydra," *J Cell Biol,* 18:367-388.

Sminth, H.M. (1955). "The perspectives of species," *Turtox New,* 33:74-76.

Smith, S.G. (1935). "Chromosome fragmentation produced by crossing-over in *Trillium erecturn L.,*" *J Genet,* 30:227.

Somani, L.L. (1995). *Dictionary of Biotechnology.* Udaipur, India: Agrotech Publ Acad, p. 496.

Sonenberg, N., Morgan, M.A., Merric, W.C., and Shatkin, A.J. (1978) . "A poly-peptide in eukaryotic initiation factors that crosslinks specifically to the 5′- terminal cap in mRNA," *Proc Natl Acad Sci USA,* 75:4843-4847.

Sonneborn, T.M. (1965a). Degeneracy of the genetic code. Extent, nature and genetic implications. In *Evolving Genes and Proteins,* eds. Bryson, V., and Vogel, S. New York: Academic Press, p. 377.

Sonneborn, T.M. (1965b). "The metagon: RNA and cytoplasmic inheritance," *Am Nat,* 99:279-307.

Spiegelman, S. and Hayashi, M. (1963). "The present status of the transfer of genetic information and its control," *Cold Spr Harb Symp,* 28:161-181.

Spirin, A.S. (1964). *Macromolecular structure of ribonucleic acids.* New York: Reinhold.

Sprague, G.F. and McKinney, H.H. (1966). "Aberrent ratio: An anamoly in maize associated with units infection," *Genetics,* 54:1287-1296.

Stebbins, G.L. (1947). "Types of polyploids: Their classification and significance," *Adv Genet,* 1:403-429.

Steitz, J.A. and Tycowski, K.T. (1995). "Small nuclear chaperons for ribosome biogenesis," *Science,* 270:1626-1627.

Stern, C. (1929). "Über die additive Wirkung multipler Allele," *Biol Zbl,* 49: 261-290.

Stern, C. and Schaeffer, E.W. (1943). "On wild-type iso-alleles in *Drosophila melanogaster,*" *Proc Natl Acad Sci USA,* 29:361-367.

Sternberg, N. (1990). "Bacteriophage P1 cloning for isolation, amplification and recovery of DNA fragments as large as 100 kilobase pairs," *Proc Natl Acad Sci USA,* 87:103.

Stout, A.B. (1917). "Fertility in *Cichorium intybus*: The sporadic occurrence of self-fertile plants among the progeny of self-sterile plants," *Am J Bot,* 4:375-395.

Stout, A.B. (1918). "Fertility in *Cichorium intybus*: Self-compatability and self-incompatability among the offspring of self-fertile lines of descent," *J Genet,* 8: 71-103.

Strasburger, E. (1877). "Über befruchtung und Zellteilung," *Jena Z naturw,* 11: 435-440.

Strasburger, E. (1882). "Über den Trilungsvorgang der Zellkerue und das Verhaltnis der Kernteilung zur Zellteilung," *Arch Mikr Anat,* 21:476-488.

Strasburger, E. (1884). *Neue Untersuchungen über den Befruchtungsvorgang bie den Phanerogamen als Grundlage für eine Theorie der Zeugung.* Jena, Germany: Fisher.

Strasburger, E. (1905). "Typische und allotypische Kernteilung. Ergebnisse und Erörterungen," *Pringsheims Jb Wiss Bot,* 42:1-17.

Streisinger, G., Edger, R.S., and Denhardt, G.H., (1964). "Chromosome structure in phage T4 I. Circularity of the linkage map," *Proc Natl Acad Sci USA,* 51:775-779.

Strugger, S. (1950). "Über den Bau der Proplastiden und Chloroplasten," *Naturwiss,* 37:166-177.

Sturtevant, A.H. (1914). "The reduplication hypothesis as applied to *Drosophila,*" *Am Nat,* 48:535-559.

Sturtevant, A.H. (1920). "The vermillion gene and gynadomorphism," *Proc Soc Exp Biol Med,* 17:70-71.

Sturtevant, A.H. (1925). "The effect of unequal crossing over at the Bar locus in *Drosophila,*" *Genetics,* 10:117-147.

Sturtevant, A.H. (1926). "A crossover reducer in *Drosophila melanogaster* due to inversion of a section of the third chromosome," *Biol Zbl*, 46:697-702.

Suderland, E.W. (1972). "Studies on mechanism of hormone action," *Science,* 177:401-408

Sueoka, N. (1961). "Variation and hetergeniety of base composition of deoxyribonucleic acids: A compilation of old and new data," *J Mol Biol,* 3:31-40.

Surani, M.A.H., Barton, S.C., and Norris, M.L. (1984). "Development of reconstituted mouse eggs suggests imprinting of the genome during development," *Nature,* 308:548-550.

Sutton, W.S. (1903). "The chromosomes in heredity," *Biol Bull Wood's Hole,* 4:213-251.

Swift, H. (1950). "The deoxyribone nucleic acid content of animal nuclei," *Physiol Zool,* 23:169-198.

Szostak, J.W. and Blackburn, E.H. (1982). "Cloning yeast telomeres on linear plasmid vectors," *Cell,* 29:245-255.

Szybalski, W., Bovre, K., Fiandt, M., Hayes, S., Hradeena, Z., Kumar, S., Lozeron, N.A., Nijkamp, H.J.J., and Stevens, W.F. (1970). "Transcriptional units and their controls in *Escherichia coli* phage λ: Operons and scriptons," *Cold Spr Harb Symp,* 35:341-353.

Täckholm, G. (1922). "Zytologische studien über die Gattung rosa," *Acta Hort Berg,* 72:297-381.

Tamkun, J.W., Schwarzbauer, J.E., and Hynes, R.O. (1984). "A single rat fibronection gene generates three different mRNAs alternative splicing of a complex exon," *Proc Natl Acad Sci USA,* 81:5140-5144.

Tartof, K.D. (1975) "Redundant genes," *Ann Rev Genet,* 9:355-385.

Taylor, J.H., Woods, P.S., and Hughes, W.L. (1957). "The organization and duplication of chromosomes as revealed by autoradiographic studies using tritium thymidine," *Proc Natl Acad Sci USA,* 43:122-128.

Taylor, W.R. (1926). "Chromosome morphology in Fritillaria, Alstroemeria, Silphium, and other genera," *Am J Bot,* 13:179-193.

Temin, H.M. (1971). "The protovirus hypothesis," *J Nat Cancer Inst,* 46:III-VIII.

Temin, H.M. and Mizutani, S. (1970). "RNA-dependent DNA polymerase in virion of Rous Sarcoma virus," *Nature,* 226:1211-1213.

Thoday, J.M. (1953). "Components of fitness," *Symp Soc Exp Biol,* 7:96-113.

Thomas, C.A. (1970). "The genetic organization of chromosomes," *Ann Rev Genet,* 5:237-256.

Thomas, R. (1955). "Recherches sur la cinetique des transformation bacteriennes," *Biochem Biophys Acta* 18:467-482.

Thompson, D.H. (1931). "The side chain theory of the structure of the gene," *Genetics,* 16:267-290.

Tischler, G. (1950). *Die Chromosomenzahlen*. Hague: Junk, W., p. 263.

Turesson, G. (1922). "The fenotypical response of the plant species to the habitat," *Hereditas,* 3:211-350.

Turesson, G. (1923). "The scope and import of genecology," *Hereditas,* 4:171-176.

Turing, A.M. (1952). "Chemical basis of morphogenesis," *Phil Trans Roy Soc B,* 237:37-72.

Uhlin, B.E. and Clark, A.J. (1981). "Overproduction of the *Escherichia coli* recA protein without stimulation of its proteolytic activity," *J Bact,* 148:386-390.

Valentine, G.G. (1836). *Repertorium für Anatomie und Physiologie, I.* Berlin, p. 120.

Van Benden, E. (1883). "Researches on the maturation of the egg and fertilization," *Arch Biol,* 4:1-4.

Vanderplank, J.E. (1963). *Disease Resistance in Plants.* New York: Academic Press, p. 206.

Varshavsky, A.J., Sundine, O.H., and Bohn, M.J. (1978). "SV40 viral mini-chromosome: Preferential exposure of the origin of replication as probed by restriction endonucleases," *Nucleic Acids Res,* 5:3469-3477.

Vavilov, N.I. (1922). "The law of homologous series in variation," *J Genet,* 12: 47-89.

Velculescu, V.E., Zhang, L., Volgelstein, B., and Kinzler, K.W. (1995). "Serial analysis of gene expression," *Science,* 270:484-487.

Veselkov, A.G., Demidov, V.V., and Frank-Kamenetskil, M.D. (1996). "PNA as a rare genome-cutter," *Nature,* 379:214.

Vogel, H.J. (1957). Repression and induction as control mechanisms of enzyme biogenesis: The adaptive formation of acetylornithinase. In *The Chemical Basis of Heredity,* eds. McElroy, W.D. and Glass, B. Baltimore: Johns Hopkins Press, p. 276.

Vogt, O. (1926). "Psychiatrisch wichtige Tatsachen der zoologisch-botanischen Systematik," *Z Ges Neurol Psychiat,* 101:805-830.

Waddington, C.H. (1932). "Experiments on the development of chick and duck embryos, cultivated *in vitro,*" *Phil Trans B,* 221:179-230.

Waddington, C.H. (1939). *An Introduction to Modern Genetics.* London: Allen and Unwin.

Waddington, C.H. (1941). "The genetic control of wing development in *Drosophila,*" *J Genet,* 41:75-139.

Waddington, C.H. (1942). "Canalization of development and the inheritance of acquired characters," *Nature,* 150:563-566.

Wahlund, S. (1928). "Zusammensetgung von Populationen und Korrelationser-scheinungen vom Standpunkt der Vererbungslehre aus betrachtet," *Hereditas,* 11: 65-77.

Waldeyer, W. (1888). "Über Karyokinese und ihre Beziehung zu den Befruchtungsvorgängen," *Arch Mikr Anat,* 32:1-35.

Walter, P. and Blobel, G. (1980). "Purification of a membrane-associated protein complex required for protein translocation across the endoplasmic reticulum," *Proc Natl Acad Sci USA,* 77:7112-7116.

Wang, A., Quigely, G.J., Kolpak, F.J., Crawford, J.L., Van Boom, J.H., Vander Marel, G., and Rich, A. (1979). "Molecular structure of a left-handed double helical DNA fragment at atomic resolution," *Nature,* 282:680-686.

Wang, J.C. (1971). "Interaction between DNA and an *Escherichia coli* protein W," *J Mol Biol,* 55:523-533.

Wang, J.C. (1985). "DNA topoisomerases," *Ann Rev BioChem,* 54:665-697.
Wang, J.C. and Liu, L.F. (1979). *Escherichia coli* DNA topoisomerase. In *Molecular Genetics,* Part III, ed. Taylor, J.X. New York: Academic Press, pp. 65-88.
Warburton, F.E. (1955). "Feedback in development and its evolutionary significance," *Am Nat,* 89:129-137.
Warmke, H.E. and Blakeslee, A.F. (1939). "Induction of simple and multiple polyploidy in *Nicotiena* by colchicine treatment," *J Hered,* 30:419-432.
Warner, J.R., Rich, A., and Hall, C.E. (1962). "Electron microscopic studies of ribosomal clusters synthesizing hemoglobin," *Science,* 138:1399-1405.
Washington, W.J. (1971). "Homoeoallelism in *Triticum aestivum,*" *Can J Genet Cytol,* 13:169-172.
Watson, J.D. and Crick, F.H.C. (1953). "Molecular structure of nucleic acids: A structure of deoxyribose nucleic acid," *Nature,* 171:737-738.
Webber, H.J. (1903). "New horticultural and agricultural terms," *Science,* 18:501-503.
Weinberg, R.A. and Penman, S.P. (1968). "Small molecular weight monodisperse nuclear RNA," *J Mol Biol,* 38:289-304.
Weinberg, W. (1908). "Über den Nachweiss des Verembung bein Memschen," *Jahresh Verein f Vaterl Naturk Würtemberg,* 64:368-382.
Weismann, A. (1883). Über die Vererbung. In *Ausfsätze über Vererbung und verwandte biologische Fragen.* Jena, Germany: Fischer, 1892, p. 73.
Weismann, A. (1885). Die Kontinuitat des Keimplasmas als Grundlage einer Theorie der Vererbung. In *Ausfsätze über Vererbung und verwandte biologische Fragen.* Jena, Germany: Fischer, 1892, p. 191.
Weismann, A. (1887). Über die Zahl der Richtungskörper and ihre. In *Bedeutung für die Vererbung.* Jena, Germany: Fischer.
Weismann, A. (1891). *Amphimixis order die Vermischung der Individuen.* Jena, Germany: Fischer.
Weismann, A. (1895). *Neue Gedanken sur Vererbungsfrage. Eine ANtwort an Herm Spencer.* Jena, Germany: Fischer.
Weismann, C., Nagata, S., Taniguchi, T., Weber, H., and Meyer, F. (1979). In *Genetic Engineering, Principles and Methods,* eds. Hollander, A. and Setlow, J.K. New York: Plenum, p. 1133.
Weiss, R.B. (1984). "Molecular model of ribosome frameshifting," *Proc Natl Acad Sci USA,* 81:5797-5801.
Welch, J.E. and Grimball, E.L. (1947). "Male sterility in the carrot," *Science,* 106:594-597.
Wells, J., Vesser, M., and Powers, D.B. (1985) . "Cassettee mutagenesis: An efficient method for generation of multiple mutations of defined sites," *Gene,* 34:315-323.
Wettstein, R.v. (1898). *Grundzüge der geographisch-morphologischen Methode der Pflanzensystematix.* Jena, Germany: Fischer.
White, M.J.D. (1935). "Röntgenbestrahlung," *Naturwiss,* 23:390.
White, M.J.D. (1945). *Animal Cytology and Evolution.* Cambridge: Cambridge University Press.

Wilson, C. and Szostak, J.W. (1995). "*In vitro* evolution of a self-alkylating ribozyme," *Nature,* 374:777-782.

Wilson, D.A. and Thomas, C.A. (1974). "Palindromes in chromosomes," *J Mol Biol,* 84:115-144.

Wilson, E.B. (1896). *The Cell in Development and Heredity,* first edition. New York: MacMillan.

Winiwarter, H.V. (1900). "Recherches sur l'ovogense et I'ovaire des Mammiféres (Lepin at Homme)," *Arch Biol Paris,* 17:33-37.

Winkler, H. (1907). "Über Pfrppfbastrade und pflanzliche chimären," *Ber Dtsch Bot Ges,* 25:568-573.

Winkler, H. (1916). "Über die experimentelle Erzeugung von Pflazen mit abrueichenden chromosomenzahlem," *Z Bot,* 8:417-53.

Winkler, H. (1920). *Verbreitung und Ursache der parthenogenese im Pflanzen- und Tierreiche.* Jena, Germany: Fischer.

Winkler, H. (1930). *Die Konversion der Gene.* Jena, Germany: Fischer.

Witkin, E.M. and Wermudsen, I.E. (1978). "Targeted and untargeted mutagenesis by various inducers of SOS functions in *Escherichia coli,*" *Cold Spr Harb Symp,* 43:881-886.

Wittung, P., Nielson, P.E., Buchardt, O., Eghlom, M., and Norden, B. (1994). "DNA-like double helix formed by peptide nucleic acid," *Nature,* 368: 561-563.

Wollman, E.L., Jacob, F., and Hayes, W. (1956). "Conjugation and genetic recombination in *Escherichia coli* K12," *Cold Spr Harb Symp,* 21:141-162.

Woltereck, R. (1909). "Weitere experimentelle Untersuchungen über Artverränderung speziell überdas Wesen quantitaltiver Artunterschiede bei Dapliniden," *Verh Dtsch Zool Ges,* 19:110.

Woodson, S.A. and Cech, T.R. (1989). "Reverse self-splicing of the Tetrahymena group I intron: Implication of the directionality of splicing for intron transposition," *Cell,* 57:335-345.

Wright, S. (1921). "Systems of mating I. The biometric relations between parent and offspring," *Genetics,* 6:111-123.

Wright, S. (1922). "Coefficients of inbreeding and relationship," *Am Nat,* 56: 330-338.

Wright, S. (1929). "Fisher's theory of dominance," *Am Nat,* 63:274-279.

Wright, S. (1931). "Evolutionary in Mendelian populations," *Genetics,* 16:97-159.

Wright, S. (1932). "The roles of mutation inbreeding, crossbreeding, and selection in evolution," *Proc Sixth Int Congr Genetics,* 1:356-366.

Wright, S. (1942). "Statistical genetics and evolution." *Bull Amer Math Soc,* 48:223-246.

Wynne-Edwards, V.D. (1962). *Animal Dispersion in Relation to Social Behaviour.* London: Oliver and Boyd.

Yanofsky, C., Helinski, D.R., and Maling, B.D. (1961). "The effect of mutation on the composition and properties of the A protein of *Escherichia coli* tryplophan synthetase," *Cold Spr Harb Symp,* 26:11-24.

Yasuda, S. and Hirota, Y. (1977). "Cloning and mapping of the replication origin of *Escherichia coli*," *Proc Natl Acad Sci USA*, 74:5458-5462.
Yates, R.A. and Pardee, A.B. (1956). "Cloning of pyrimidine biosynthesis in *Escherichia coli* by a feed-back mechanism," *J Biol Chem*, 221:757.
Young, N.D. and Tanksley, S.D. (1989). "Restriction fragment length polymorphism maps and the concept of graphical genotypes," *Theor Appl Genet*, 77:95-101.
Zaenen, I., Larebeke, N., Teuchy, H., and Van Montagu, M. (1974). "Supercoiled circular DNA in crown gall inducing *Agrobacterium* strains," *J Mol Biol*, 86:109.
Zarrinker, P.P. and Williamson, J.R. (1994). "Kinetic intermediates in RNA folding," *Science*, 265:918-924.
Zechel, K., Bouché, J.P., and Korenberg, A. (1975). "Replication of phage G4.A novel and simple system for the initiation of deoxyribonucleic acid synthesis," *J Biol Chem*, 250:4684-4689.
Zimmer, E.A., Martin, S.L., Beverly, S.M., Kan, Y.W., and Wilson, A.C. (1980). "Rapid duplication and loss of genes coding for the α chains of hemoglobin," *Proc Natl Acad Sci USA*, 77:2158-2162.
Zinder, N.D. and Lederberg, J. (1952). "Genetic exchange in *Salmonella*," *J Bact*, 64:679.
Zissler, Y. (1967). "Integration-negative (int) mutants of phage λ," *Virology*, 31:189.
Zubay, G., Schwartz, D., and Beckwith, J. (1970). "Mechanism of activation of catabolite-sensitive genes: A positive control system," *Proc Natl Acad Sci USA*, 66:104-110.
Zukerbandl, E. and Pauling, L. (1965). "Evolutionary divergence and convergence in proteins," In *Evolving Genes and Proteins*, eds. Bryson, V. and Vogel, H.J. New York: Academic Press, pp. 97-166.
Zukov, N.I. (1941). "Alteration of the nature of plants by means of interspecific hybridization in the genus Nicotiana. (Russian) *Vsesojunznyi tabacn i machorocx, Promyslennosch (Krasnodar)*, 143:142.